Landslips

THE AXMOUTH TO LYME REGIS UNDERCLIFFS NATIONAL NATURE RESERVE

Donald Campbell

Also by Donald Campbell

The Encyclopedia of British Birds (1999) - A Dempsey Parr book for Paragon, 384pp.

Exploring the Undercliffs (2006) - Coastal Publishing, for the Jurassic Coast Trust and English Nature, 64pp.

Rocks and Wildlife around the Axe (2020) - Axe Vale & District Conservation Society, 85pp.

Landslips

THE AXMOUTH TO LYME REGIS UNDERCLIFFS NATIONAL NATURE RESERVE

Donald Campbell

Axe Vale & District Conservation Society

First published in 2023 and © by Axe Vale & District Conservation Society
Ruth Gray, Treasurer and Membership Secretary of AVDCS
www.axevaleconservation.org.uk
Text © Donald Campbell 2023

ISBN 978-1-9162819-1-2

The right of Donald Campbell to be identified as the author of this work has been asserted in accordance with the Copyright, Designs and Patents Act 1988, sections 77 and 78.

No part of this publication may be reproduced, stored in a retrieval system or transmitted in any form or by any means without the prior permission of the publisher and copyright owner.

This book is sold subject to the condition that all designs are copyright and are not for commercial reproduction without the permission of the designer and copyright owner.

Photographs and other images are credited at the end of the book. All unmarked photographs were taken by the author. Whilst every effort has been made to obtain permission from the copyright holders for all material used in this book, the publishers will be pleased to hear from anyone who has not been appropriately acknowledged, and to make a correction in future reprints.

Designer & Editor: John Marriage

With thanks to the organisations that have supported the production of this book:

This book is dedicated to my five grandchildren who share my interests in wild places and wild life. Hamish and Tilly live beside the estuary of the Morar river, looking over to Skye and the Small Isles, while Angus, Iona and Freya, in Buxton, are well placed for the Peak District. May the changing climate not impinge too much on their lives.

Contents

	Foreword	i
	Introduction	vi
1	Undercliff history and exploration since 1630	1
2	Early Days of the National Nature Reserve	25
3	The great landslip of 1839 and subsequent changes	43
4	More landslides and some investigations	69
5	The first 40 years of the National Nature Reserve	85
6	Norman Barns, early surveys and the first two management plans	101
7	Birds and other animals 1950 to 2003	121
8	Soft cliffs and managed grasslands	143
9	Plant and animal diversity 2000 to 2015	163
10	Flies, moths, ants and measuring tree trunks	189
11	Exploring the cliffs, the shore and life in the sea	201
12	Some achievements and a long look back to very different places	223
	Bibliography	245
	Index	251
	Image Credits	261

Data Summaries and Species Tables

Charts 1 & 2	Rainfall records 1870-2019	73
Table 1	Major landslips, 1876 - 2014	76
Table 2	Important features of the reserve (second management plan 1992)	94
Table 3	Relative abundance and trunk size of the larger trees (Hamish Archibald in the 1965 management plan and Donald Campbell in 2021)	105
Table 4	Moss species found in Culverhole in July 1995 (W and J Cox)	109
Table 5	Some agarics of interest, 1988-2013 (David Allen)	110
Table 6	Woodlouse families in different habitats (UCL conservation group 1966)	127
Table 7	Terrestrial molluscs (Norman Barns 1990)	132
Table 8	Millipedes (Ade Turner 1990)	133
Table 9	Breeding birds 1994-98 (Donald Campbell)	135
Table 10	Bees and wasps (Tom Wallace 1961 and Mike Edwards 1995)	138
Table 11	Crickets and grasshoppers (Tom Wallace 1961 and Mike Edwards 1995)	139
Table 12	Nationally scarce and Red Data Book invertebrates (Mike Edwards 1995 and David Gibbs 2003)	140
Table 13	Total species and number of Red Data Book invertebrates found in five locations in 2002 and 2003 (David Gibbs)	172
Chart 3	The number of RDB, scarce and other invertebrates at four soft cliff sites. (David Gibbs)	173
Table 14	The most common birds in winter and early summer in the 2007-2011 BTO survey (Donald Campbell)	176
Table 15	The most recorded butterflies 2001- 2010 (Phil Parr)	177
Table 16	Butterfly flight periods (Phil Parr)	178
Table 17	Classification of land animals found (Natural England Bioblitz 2011)	185-6
Table 18	Summary of plants and their allies (Natural England Bioblitz 2011)	186
Table 19	Red Data Book flies (Martin Drake 2017)	191
Table 20	The most frequently recorded fly species (Martin Drake 2020)	191
Table 21	Zonation of plants and animals along three Undercliff beaches (Ambios Environmental Consultants 1995)	211
Table 22	Bioblitz sponges (Devon Sea-search 2011)	214
Table 23	Sea anemones in Charton Bay (Devon Sea-search 2011)	215
Chart 4	Insecticide residues in the breast muscle of different bird species (Colin Walker)	241

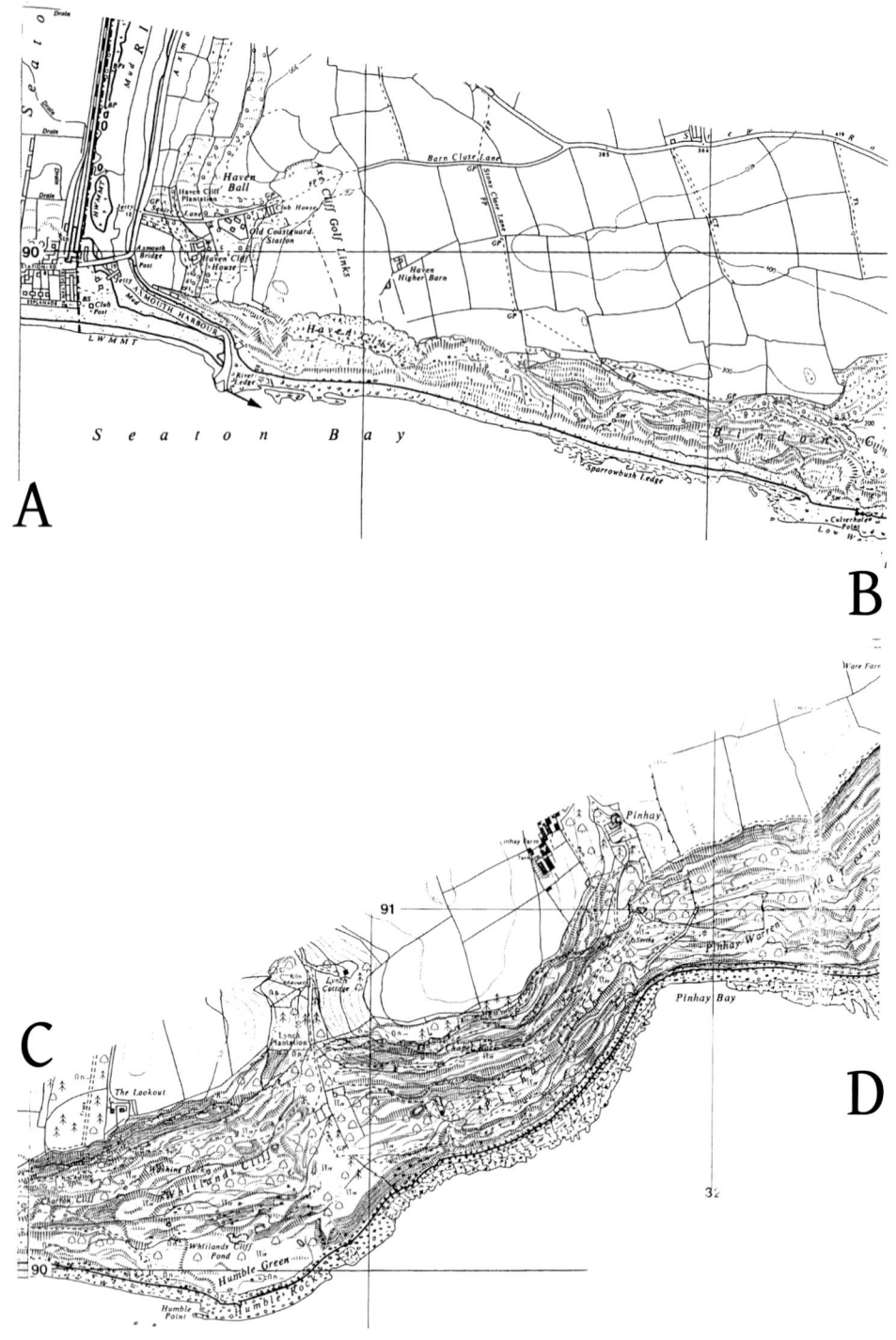

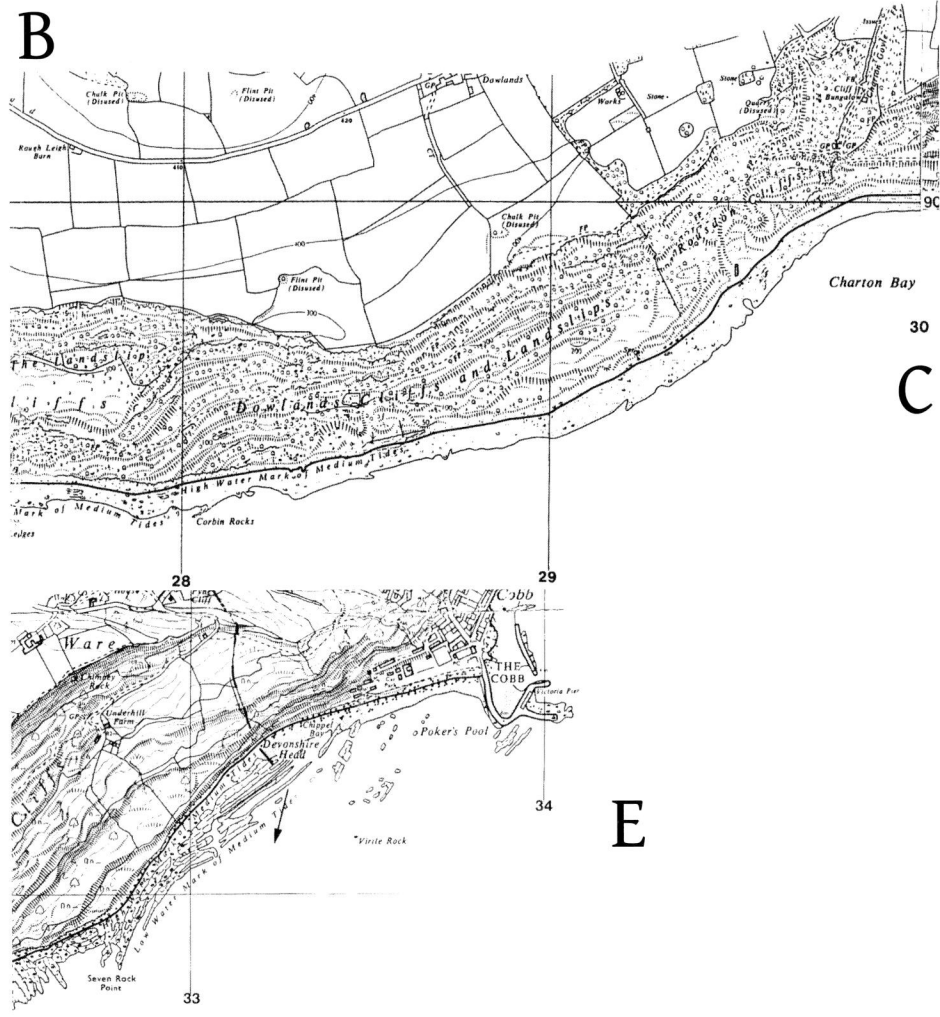

Map of the Undercliff, running from Axmouth Harbour in the west (A) to Lyme Regis Cobb in the east (E). The sections link up in alphabetical order.

Foreword

Unique is a much-abused word but there is really no other term by which to describe the six miles of coastal landslip complex known as the Undercliff. It is the only place where rocks from all three stages of the Mesozoic era are bought together cheek by jowl by circumstances of geological caprice. The complex geology and the perpetual state of landslipping are the crucial elements that form and influence the array of habitat types and successional stages of this self-sustaining and ever-changing wilderness landscape.

Donald has been exploring the Undercliff for nearly 30 years. Exploring is not an overstatement as so much of this six-mile wilderness is scarcely accessible unless you have the determination to discover and break new ground as Donald has done.

The Landslip (Landslip and Undercliff are used interchangeably in the local parlance) will often exact a tithe from the explorer, usually paid in blood and collected by some grasping briar or razor flint. For the geologist, the naturalist and the historian, the price is worth paying as it is only by exploring with some purpose in mind that one can start to truly understand the chaotic essence of this strange place and the processes that form it.

This may be England's most dangerous nature reserve. Away from the path natural hazards are frequently encountered including, but not limited to, high cliffs, deep subterranean chasms, scree slopes, rock falls, quicksands and mud flows. Accident and misadventure await those who underestimate the difficulties of the terrain or fail to treat the place with the respect it deserves. Getting lost is a possibility for the seasoned explorer. For the uninitiated it becomes the likelihood.

There are really very few places in these islands where, by forcing ones way through clawing bramble thickets, rounding precipitous ivy clad crags and negotiating a chaos of gullies, ridges and scree slopes, you may arrive at a view that no one else has enjoyed for 50 years or a century or more. We must imagine that Donald will have enjoyed many such views over the years. This is a truly wild landscape, described by Professor Denys Brunsden as the finest wilderness area in Southern Britain, though that was not always the case. In this book Donald shares what he has discovered about a landscape that does not yield her secrets easily. We must thank him for taking the not inconsiderable time required to understand the essential character of this fabulous, sometimes overwhelming and unruly tangle of coastal landslip where change is the only constant.

In 2006 Donald published his book *Exploring the Undercliffs* and had, by then, already more than earned the right to title it so. The book was published to commemorate the 50th anniversary of the creation of a National Nature Reserve. This book expands on many of the themes of that first publication on the subject and contextualises those themes more fully.

For the historian this text includes some of the earliest 18th century antiquarian accounts of the Undercliff. We must surmise that the Landslip was well known to earlier cottagers, stockmen and fishermen but no written evidence has come to light. Donald documents the beginnings of the era of tourism which largely evolved in the area after the events of the great landslip of 1839 and details the working landscape of the Landslip prior to that event when stock roamed over an essentially open landscape where cottagers cut coppice and kept orchards and market gardens.

The events immediately after the great landslip drew early men of reason to the area to develop their theories in the fledgling science of geology. The Undercliff has continued to attract geologists and palaeontologists ever since.

Agricultural activity seems to have ended in the second half of the 19th century but the Undercliff wasn't totally abandoned. Small scale quarrying and lime burning continued. Water extraction became a local industry that continued into the 1990s and there were some unsuccessful experiments with forestry and some economically painful attempts to extract some of the well grown Undercliff timber in the 20th century. What was once a managed and open landscape populated by smallholders had become thick, tangled woodland by the second half of the century.

A little-known historical aspect of the Undercliff concerns the establishment of the National Nature Reserve (NNR) and the characters involved in the process. The Undercliff was not considered in the first tranche of NNRs in the wake of the post-war National Parks and Access to the Countryside Act of 1949. Finally declared a National Nature Reserve in 1955, the Undercliff NNR evolved directly out of pivotal legislation that placed legal commitments to nature conservation on the statute books of British law for the first time. Today, the Undercliff is one of the most highly protected and designated sites in the country, being a Site of Special Scientific Interest, National Nature Reserve and Special Area of Conservation within an Area of Outstanding Natural Beauty and, since 2001, part of a World Heritage Site.

These accounts are more than a timeline. Within them we can chart a change in social attitudes and popular interpretations of wild places and wild things. It is perhaps timely now, with the rewilding movement gathering influence, to reflect on the Undercliff as possibly one of the original rewilded landscapes. At Goat Island, where fields that were wheat and turnips in 1839, or at The Chasm that formed behind in the same huge landslide event, we now see rich, diverse mature woodland and scrub. This is the result of a largely non-interventionist rewilding approach, and the steady influence of a classic ecological succession. Rewilding has created the wilderness we now see.

The naturalist will not be left wanting from this book. It is a most thorough account of the ecology, habitat types and associated species found on the reserve. Donald has been the hub of the Undercliff for many years and by consolidating the rather disparate but extensive reserve records and drawing on the knowledge of his numerous friends and associates, he has produced the definitive natural history of this stretch of coast. Taxonomic groups are discussed in detail as are the range of habitats, along with the natural processes that form and sustain them.

Little will be understood about any aspect of the Undercliff without a foundation in the geomorphological processes that ultimately form and regulate the place. Donald takes us through a history of the scientific accounts of the land slip process from early pioneers in the field up to the very latest theories and geological investigations. We learn about the various landslip mechanisms involved in producing earth movements of the constant and gradual kind along with the failures that have happened on an altogether grander scale.

The geodiversity of the site, a result of the great unconformity, is well documented within these pages and the palaeontological interest, for which the Undercliff is so renowned is also covered in detail. The major fossil groups and their parent rocks are described. The site is world-renowned for its lower Jurassic marine reptile assemblage and has yielded type specimens for 15 species. It is the type locality for the Lias series in Britain and has important exposures of upper Triassic and upper Cretaceous sequences, having also yielded type specimens of a rare upper Cenomanian ammonite assemblage.

Finally, the author guides us through the management of the site as a nature reserve, an area of work in which he has been a very active participant and dedicated advocate over the years. Active management involves protecting the geological and biological interest features of the site. Habitat management work helps to maintain some of the few remaining areas of open, species rich grassland as well as controlling the extent of non-native species. Public access provision is also a challenging management issue for such a mobile and dynamic site. Donald brings his thoughts to bear on the changes and developments, both good and bad, that the future may bring.

The Undercliff is ultimately much more than a geological oddity, a unique place or a nature reserve. It is the focus for a motley community of geologists, hikers, explorers, survivalists, surfers, naturalists, clandestine horticulturalists, palaeontologists, nudists, local historians, conservationists, volunteers, fossil collectors, fishermen, beachcombers, contractors, artists, musicians and writers, each of whom have a stake in the place, a unique insight and a page to write in the story of what the Undercliff means today.

From sourcing original research material to gathering extensive interview excerpts, Donald has lavished care and attention on his subject. From the man who knows this place better than anyone, this book tells the Undercliff story in an engaging and approachable style and is recommended to anyone with a love for wild country and wild things.

Tantum mutatio est permanens.

Tom Sunderland and Rob Beard

Reserve Managers, Axmouth to Lyme Regis Undercliffs
National Nature Reserve.
January 2022

Raised reefs occasioned by the Landslip.
Drawn and published by J Baker.

Introduction

Soon after Nicky and I moved to Devon in 1993 I wanted a project which could help me to get to know the area better. When I approached John Woodland, the county's representative for the BTO, British Trust for Ornithology, he suggested that a bird census in the Undercliff would be useful and that I should contact Norman Barns, the local authority on the area. Norman immediately took me to Goat Island and the Plateau and introduced me to the local conservation society of which he was President.

After five years of census work in the Undercliff I knew much of the area reasonably well and had become Chairman of the Society which I hoped would become more active in practical ways. This led to contact with Albert Knott, English Nature's manager in the Undercliff. Soon the Society was helping on the Reserve and in 2000 Albert asked me to create a reference resource about the area.

It was soon obvious that English Nature's files, their Management Plans and Geological Reports together with the resources of the Philpot Museum in Lyme Regis would keep me busy for months, but it was equally clear that I needed to talk with some of those who knew different aspects of the Undercliff. Norman Barns and Tom Wallace, who had both contributed extensively to knowledge of its plants and animals, were the first to be interviewed. John Fowles was next as his book *The French Lieutenant's Woman* was partly set in the Undercliff which John had known well when he lived at Underhill Farm at the eastern end of the Reserve. Later the book was made into a very successful film, much of it shot locally in woodland owned by the Allhusen family. I therefore contacted Dracaena, the widow of Ormsby Allhusen, who had been High Sheriff of Devon and involved in the establishment of the Reserve. She lived immediately above the Undercliff which she knew well and in which she often walked. My knowledge was increasing, but the gaps in it were also becoming clear and experts in some of the less familiar taxonomic groups were needed. Soon enthusiasts for mosses, fungi and lichens were invited to come, to explore, make species lists and report back. By the end of 2001, when the coast became a World Heritage Site, I had produced a long handwritten account which was made readable by Nicky's daughter Sarah who had managed to interpret my messy scrawl and to cope with the many scientific names of plants and animals.

In June 2000 the case for World Heritage Status had been put forward in the Nomination Book to which Prof Denys Brunsden had been a major contributor. As Chairman of the Dorset Coast Forum he had *"enthused the World Heritage Nomination process over the last five years"*. Despite this deep involvement in the promotion and protection of the coast he found time to comment helpfully and extensively on the early parts of my account. Later he suggested that I should write a 50th anniversary guide to the Undercliff as part of a series on aspects of the World Heritage Coast. With this in mind he and Albert came to our home in Combpyne to begin to plan what would become *Exploring the Undercliff*. With their support and with help from Peter Sills and Jonathan Lewis of Coastal Publishing, the book was launched in Lyme Regis in February 2006.

Its publication made me keen to write another book but Nicky was more quickly organised and her *To Buy a Whole Parish* was published in 2015. She describes how Sir Henry Peek established the Rousdon estate and the stern but fair way he treated his workers and their children. I was still struggling to link text and the many potential illustrations I had collected until eventually John Marriage agreed to take on the task of book production. In 2020 *Rocks and Wildlife around the Axe* was published by the Conservation Society.

Immediately after it was launched at Seaton Jurassic, Covid-19 led to lockdown. While this made it extremely difficult to sell books it also provided plenty of time for writing another one. The Undercliff was an obvious topic. With much new material since the earlier book there was no shortage of subject matter and this was substantially increased when I decided to refer back almost 500 years to John Leland and his undigested notes on *The antiquities of this nation*. John Marriage has again played the major role in its production and getting Out of Hours Typing to interpret my handwriting.

Twenty years earlier Ken Gollop, Colin Dawes and Chris Davis had been among those who had added to my knowledge of sea life, and their contributions now form part of Chapter 11, "Along the shore and out to sea". The final chapter includes some recent thoughts from David Allen, Rob Beard and Tom Sunderland.

Tom and Rob have been kind enough to write a foreword mentioning my contributions to recording and at times working to maintain and improve access and grassland. They have also commented on the text as have David Allen, Jane Daunsey and Richard Matthews (Woody). Mike Lock, Chairman of AVDCS and an experienced editor, has greatly improved punctuation, checked scientific names and corrected a couple of errors.

As mentioned many images came long ago from Jo Draper at the Philpot Museum while the files of Natural England and English Nature have been equally valuable, as have the photographs of Paul Naylor, Phil Parr and Colin Varndell. The Field Studies Council, the John Muir Trust and the Grover family archives have provided additional material. Their illustrations and others from a range of sources are individually acknowledged on pages 261ff.

CHAPTER ONE
History and Exploration

"The Axe flows into the sea through a trough washed out of the blood red sandstone that comes to the surface between the hills of chalk... On the further side, that to the rising sun, the chalk with dusty sandstone underneath, rears itself into a bold headland, Haven Ball, that stands precipitously against the sea, as a white cold shoulder exposed to it. Up a hollow of this hill, a combe as it is called, a mean track ascends to the downs which overhang the sea and extend, partly in open tracks, in part enclosed, as far as Lyme Regis."

Novelist, landowner, priest and hymn writer, Sabine Baring-Gould, described the western extremity of the Undercliff in his 1899 book, *Winefred*. The story was set much earlier, in the Napoleonic Wars, when smuggling on the Devon coast was widely practised and often accepted. Winefred and her mother had lost their home in a landslip and, having failed to get work in Seaton, set out on a series of adventures involving smuggling, chance meetings, climbs and cliff falls, before a final dramatic landslip engulfs the treacherous, unscrupulous boatman who had brought them across the Axe.

A mile upstream Axmouth had been a thriving port, and there had been plenty of descriptions of the river and surrounding area even before 1546 when John Leland, a great traveller with a deep interest in ancient manuscripts, visited Seaton. Despite his extensive explorations searching monastic and collegiate libraries with the intention of writing *The History and Antiquities of this Nation*, he had only written a mass of undigested notes. 150 years later, Thomas Hearne's version of *The Itinerary of John Leland the Antiquarry* appeared in nine volumes between 1710 and 1712. Later still, in 1838, this source was used by D M Stirling in his *Guide to the Watering Places on the South East Coast of Devon*.

"There hath beene a very notable haven at Seton, but ther lyeth

Chapter 1

between 2 pointes of the old haven a mighty rigge and barre of peble stones in the very mouth of it: and the river of Ax is driven to the very est point of the haven called Whit-cliff, and ther at a very smaul gut goeth unto the se, and her cum in small fischar boats of socour. The town of Seton is now a mean thing, inhabited by fischarmen".

Stirling also quotes a writer in 1630 *"Axmouth lieth on the east side of the River Axe, where if poureth itself into the sea, from whence it hath its name. The place is a large fair bay, and hath in former times yielded good harbour to ships lost in tempestuous weather. It appeareth that in this place, divers works have been attempted, for the repairing of the old decayed haven, but of late years with better success than formerly, by T. Erle Esq., lord of the land; who, when he had brought the same to some likelihood, was taken away by death, leaving his labours to the unruly ocean which together with unkind neighbours (by carrying away the stones of that work) made a great ruin of that work".*

Haven Cliff. about 1948

His son attempted to repair what was left but the attempt failed within ten years. In 1806, recovery was, again, proposed but little was done until John Hallett *"completed a commodious harbour in the mouth of the Axe. Extensive coal, culm and timber yards have been erected on both sides of the river"*. However, like the other ventures, success was short lived and the coming of the railway made other projects, except the concrete bridge over the river, unlikely.

Francis Kilvert, the writer of well-known diaries, was on Seaton beach in 1873 just after an excursion train had come in. *"The steep streets of the bright little town were busy with people moving up and down. The beach was thronged, swarming, a gay merry scene, light dresses, parasols, straw hats and puggarees, lovers sitting under the shade of boats, unloved girls looking on at undisguised blandishments, and girls with shoes, stockings and drawers off*

2

History and Exploration

wading in the tide, holding up their clothes nearly to their waists and naked from the waist down."

Through misadventure, Kilvert was even more naked *"a boy brought me to the bathing (machine) door two towels, as I thought, but when I came out of the water and began to use them, I found that one of the rags he had given me was a pair of very short red and white striped drawers to cover my nakedness. Unaccustomed to*

The Mouth of the Axe, 2018

such things and customs, I had in my ignorance bathed naked and set at nought the conventionalities of the place and scandalized the beach. However, some little boys who were looking at the nude naked man appeared to be much interested in the spectacle and the young ladies who were strolling by had no objection".

Soon a much more dramatic distraction would enliven the beach for a heavily laden collier had come into the bay for the coal to be unloaded into barges and then into carts. *'When a coal cart was full, six horses, in a long string, were hitched to it and the cart was dragged up the steep shingle bank and ploughed deeply into the loose shale and rattling pebbles. When the cart had been emptied at the wharf, the shaft horse was sent galloping down the beach going full shift*

Haven Cliff, 2018

through the throngs of holiday folk... followed by a string of five horses at a long trot or cantering, pebbles flying right and left and the chains rattling like the bottomless pit.'

Perhaps the excitements of the day and, the prospects of the train journey home, kept people on the beach or else the view of the

Chapter 1

steepness of Haven Cliff put them off a potentially arduous climb, but whatever the reason, few entered the Undercliff from this direction. By contrast, Lyme had long been popular and its links with the picturesque movement had encouraged exploration.

At the time of Hearne's account of Leland's itinerary, Lyme was losing its importance as a port but it was not long before its decline began to be reversed. William Gilpin's promotion of the picturesque helped this recovery. The discovery of romantic landscape and wild nature, together with the new popularity of sea bathing brought the right people to the town.

An inherited fortune allowed the Rev John Swete to indulge his passion for the picturesque as he travelled around Devon in 1794, painting and describing his experiences in a million surviving words. Todd Gray, editor of *Travels in Georgian Devon* (1998), comments on Swete's preoccupation with the 'picturesque' for he used the word 142 times in volume 2 of *Travels* compared with the use of 'beautiful' 73 times and 'romantic' 47 times but with a mere four mentions of the 'sublime'. At first Swete thought Lyme *"conspicuously beautiful"* being *"overhung by mountainous height"* but soon his small party were driven away *"by the dirtiness of the inn, the badness of the cooking and the impertinent incivilities of the people who kept it."* Lyme was *"not an eligible place for summer resort. It was void of trees, had no picturesque views, no rides, no diversity of walks – the shore was pebbly, neither well adapted for bathing, exercise or amusements."*

Leaving Lyme on the old road he thought that Pinhay House *"had a degree of respectability"* and as it was surrounded by a number of trees *"possessed a greater attraction."* On reaching Seaton, just beginning as a resort, he found more problems for there was *"but little extent of beach, and that bad, yet between high mountainous hills it had a sheltered situation and the environs, in point of picturesque beauty and varied rides are equal, if not superior to most of the coast."*

At much the same time Jane Austen, working on the first draft of *Sense and Sensibility*, feared that the picturesque had been trapped in its own jargon. *"You must not enquire too far Marianne – remember I have no knowledge of the picturesque and I shall offend you by my ignorance and want of taste, if we come to particulars. I shall call a hill steep which ought to be bold, surfaces strange and uncouth which ought to be irregular, and rugged and distant objects out of sight which ought only to be indistinct through the soft medium of a hazy atmosphere."*

Slightly earlier in 1772, William Pitt, Earl of Chatham had spent a few days at The Three Cups and came again, with his two sons, the next year. The younger, fourteen years old, also called William, organised family excursions and was all for visiting *"the greatest wonder in nature, a petrifying spring at Whitlands. His ardour led them rather far in the Whitlands excursion and as all Lyme people know, walking there is none of the easiest. Next day, the elder Pitt was reminded of his gout - legs mend, he said, but slowly."* Gout and advancing years were amongst his disorders. He also wrote of *"breathing the purest air imaginable, pursuing health through paths of amusement over the hills which abound with striking beauties of nature."*

Ten years later wealthy landowner Copplestone Warre Bampfylde, from Hestercombe in Somerset, spent a week in the town painting some of the local views which included two of his favourites, both set in the romantic Undercliff. One was Whitlands Cliff, which was then largely open, rocky grassland with a scattering of trees, the other was the Humble Rock nearer to Lyme Regis. His painting of the rock shows an open view towards Golden Cap and the Dorset coast and includes a man with a gun and a spaniel, as well as figures clambering up from the beach. Today the Humble Rock is so obscured by trees and ivy that is difficult to find. When Hestercombe was sold in 1872 several albums of Bampfylde's watercolours went with it; our two were in a collection of 120 watercolours and monochrome drawings which are now in the

Whitlands Cliff

Chapter 1

The Umble (or Humble) Rock

Victoria and Albert Museum.

George Roberts (1804 to 1860), often described as the first historian of Lyme Regis was, not surprisingly, more positive about the town than Swete but he was even more enthusiastic about Pinney cliffs. The owner Mrs Edge "obligingly allowed parties to visit, but not on Sundays, once they had asked permission." Once there Roberts described how *"a deviation from the path, in order to attain different points of view, exposes one to inequalities of a fatiguing character but every step is on romantic ground – new embellishments, features and combinations, continually rise into view causing a rapture that almost renders one insensible to fatigue"*.

Walter White who worked in the library of the Royal Society had more right to feel fatigue as he walked from London to Lands' End in 1855. Much of his description after walking into the Undercliff still applies *"The hills here rise to a height of about five hundred feet, in huge extended masses of Chalk and Greensand resting on Lias and Red Marl, a formation more than usually liable to disturbances from the weather; for after abundant rains the two upper deposits, loosened by the percolation of water, slide away from the lower two by whole acres at a time; sinking here into hollows and pits, there a ridge leaning inwards, yonder a shelf like a great step and all so broken up with steep banks, hummocks and knolls, as to form a very chaos. Imagine all this, when after a lapse of years, the perpendicular wall behind is faced with foliage; when the rugged slope heading down to the shore is covered in Copse wood; when the hillocky shelf midway is carpeted with the softest turf, its deformities beautified or concealed by luxuriant vegetation and you will have an idea of the Undercliff which stretches nearly the whole distance from Lyme to the mouth of the Axe"*.

History and Exploration

"And here you may wander at will: up and down and in and out among the grassy knolls and flower thickets... now treading your way among Foxgloves so tall as to bring their dappled bells to a level with your eye; now doubling a dense bed of Thistles, Nettles or Gorse, cumberers of the soil in other places, but here playing an effective though subordinate part in the general luxuriance. Yonder a grey, old, ivy-coated turret projects from the screen of wood on the cliff above; coming nearer, you find it to be a buttress of limestone left standing when the chalk fell away; and beyond it more of the grey crags and red gravel peeping out from the abounding foliage".

White felt no need for a guide as *"you have the sea on the right or left according as you are journeying east or west and the high cliffs on the other hand, you cannot fail of arriving in time at either extremity"*. He arrived in due course, at the western extremity, and walked on to Branscombe where he spent the night at the Mason's Arms. He found it *"comfortable enough although the hostess saw fit to apologise for the rustic nature of the accommodation"*.

Two years after White's walk, Roland Brown's *Beauties of Lyme Regis and Charmouth* was published. He thanked *"the indefatigable and esteemed"* George Roberts and was delighted, as Walter White had been, that he could do without a guide and also without money for *"no fee is demanded before admission is obtained and no restrictions interfere with our liberty or natural love of enjoyment."*

Brown *"anticipated a time when, aided by steam communication, its various attractions will be so well appreciated that it will occupy that prominent position on the list of watering places to which its superior qualifications entitle it"*.

Lyme Church and the Cobb

He was soon writing of some of those qualifications as standing in the pathway that opened before him *"we find that we are surrounded, on*

Chapter 1

every side, with beauties scattered by Nature in her wildest exuberance of joy and after proceeding a short distance, we pause to survey the most prominent objects. Up the side of the cliff, we may trace several serpentine paths winding under the green boughs of some favourable eminence. One of these walks conducts us to Chimney Rock, which, seen at a distance, presents the appearance of a weather-beaten turret of some romantic Castle surrounded by its deserted gardens, where all is left to grow just as it likes. Upon drawing near the summit, however, we find that it is no artificial structure, but a ponderous crag, jutting out from the cliff, to which ivy and other creepers tenaciously cling".

Chimney Rock and Underhill Farm, about 1910.

"The ramble to this point is particularly delightful and well repays the fatigue of ascent, for when we have reached this height, we obtain an extensive view of the Undercliff, a chaotic mass of hills, steep banks and knolls, here and there separated by fertile surfaces of pasturage where flocks of sheep are quietly browsing surrounded by little clumps of trees and luxuriant thickets!"

John Fowles' novel *The French Lieutenant's Woman* is set in 1869, 12 years after Brown's excursion. In Chapter ten, Charles Smithson, wealthy baronet with a taste for fossils and young women, sets out along the beach. He soon found *"a fine fragment of Lias with Ammonite impressions, exquisitely clear, whirled galaxies that Catherine-wheeled their way across ten inches of rock"*. A fine gift for Ernestina, a possible wife. Charles, who liked to think of himself as a scientific young man, had become increasingly interested in palaeontology specialising in echinoderms, or petrified sea urchins, sometimes called tests. These do not come out of the Blue Lias, but from the superimposed strata of flints, so Charles, after an ignominious fall among the boulders of the beach, climbed up a steep but safe path which led up the cliff to the dense woods above.

"Those woods are on a mile long slope caused by the erosion of the ancient vertical cliff face... this steepness, in effect, tilts it and its vegetation, towards the sun; and it is this fact, together with the

water from countless springs, that have caused the erosion that lends the area its botanical strangeness – its wild arbutus and ilex and other trees rarely seen growing in England; its enormous ashes and beeches; its green Brazilian chasms choked with Ivy and the liana of wild Clematis, its bracken that grows seven, eight feet tall; its flowers that bloom a month earlier than anywhere else in the district. In the summer it is the nearest this country can offer to a tropical jungle. It has also, like all land that has never been worked or lived on by man, its mysteries, its shadows, its dangers – only too literal ones geologically since there are crevices and sudden falls that can bring disaster, and in places where a man with a broken leg could shout all week and not be heard".

In the woods, Charles, having disturbed Sarah, the French Lieutenant's Woman, was now heading east *"putting his best foot forward and, thoughts of the mysterious woman behind him, he walked a mile or more, until he came simultaneously to a break in the trees and the first outpost of civilisation. This was a long, thatched cottage which stood slightly below his path. There were two or three meadows round it running down to the cliffs."*

This is where John Fowles wrote the book, later made into a successful film, which introduced many people to the Undercliff. In April 2001, as part of my Undercliff research for English Nature, I interviewed John at his home in Lyme Regis. Just as the novel showed his knowledge of the local area, so his talk reflected his deep interest in the places he had explored before his walking became restricted.

Partly because of his long association with the Philpot Museum, he was delighted that *"the UNESCO people"* were going to declare the coast a World Heritage Site. It was also because *"it will give the whole economy a much needed jolt"*. He recalled a number of distinguished geologists and palaeontologists including Chris McGowan from Toronto, a specialist on Ichthyosaurs *"who had just done an excellent book"*. He recalled geologist Muriel Arber *"a marvellous example of Cambridge who never said anything bad"*. She had covered all the coastal areas *"wherever water oozes out"*. She was due to arrive the next day on one of her regular visits. *"I must take her to see my gardener at Underhill Farm"*. John had met, and very much liked, Elaine Franks, when she was working on her *Sketchbook of the Undercliff*. *"It was extraordinary how short sighted she was, she had great difficulty, could not see the birds and butterflies' wings. In a sense it was remarkable, it encouraged me that she had such sharp eyes!"* Linked with discussion of his favourite Undercliff prints, he talked of

Chapter 1

Field pattern at Underhill Farm in the 1950s

Nigel Cozens' *"very expert, a kind of dream, a good book seller. I suppose I like peculiar parts of the book trade and Nigel provides them!"*

John remembered voluntary warden, Tom Wallace, who had often parked at Underhill Farm, and he had sent Tom bird records, such as the Hen Harriers which he had seen two or three times coming in from the sea. After mentioning Ravens and Hornets, he spoke of other favourites like the Blue Fleabane and the Strawberry Tree which he had grown from Undercliff seed. Above all, he enjoyed Bee Orchids and Marsh Helleborines, and explained the range of his biological interests with talk of *Cicindella germanica*, a rare Tiger Beetle, recently found near Bridport.

At the time of the novel, Ware Common extended beyond the farm and a 1904 photograph shows trees to the right, but open fields on the left. If in 2001 you entered the Undercliff over the stile, which has now gone, you were among Sycamores and Bracken. The encroachment of Bracken was mentioned by Norman Barns in a series of notes written in 1985 giving guidance to any potential walk leaders. Norman also mentioned the steps among the minor landslips which occurred between 1961 and 1964 before the path reached Ravine Pond. This Duckweed-covered pond, formed in 1961, is fed by small springs and in 2003 it smelled particularly unpleasant when English Nature and the Axe Vale Conservation Society volunteers removed many of the fallen trees and branches and felled and cleared a number of others nearby to let in more light.

On the seaward side of the pond are the remains of walls built for John Ames who had settled at Pinhay in 1843. The walls are mentioned by Rowland Brown when he took the path *"from which many others deviate; proceeding through an unbroken piece of ground known as Donkey's Green, towards Pinhay, distinguished by the Whitechapel Cliffs and heedless of the barriers placed to impede our progress, pass down the dull, ill natured, envious looking path which conceals for a few minutes the surrounding beauties of Pinhay Cliffs. We do not wish to open afresh an old sore; therefore, nothing shall be said of those unpleasant circumstances which the walls on either side suggest; suffice it is to observe that the narrow passage well repays the fatigue of descent, for after the walls are discontinued, through the gaps in the fences, we obtain some of the most charming views of the sea and the hills".*

Following tree growth and landslipping, these views are now both different and restricted, but the area of Donkey's Green or Pinhay Warren is a classic 'soft cliff' site where many unusual invertebrates have been found.

Underhill Farm around 1960

Norman Barns's notes go on to describe a cast-iron ram-housing with water issuing from pipes. This was the water supply for Pinhay House. He then describes the Pinhay fault with the Blue Lias to the east and White Lias and Greensand to the west with a mud flow moving like a glacier in between. By the track from Pinhay House, a carriage drive before its use as access to the Whitlands pumping station, is a huge Small-leaved Lime with a 'V' formed double trunk, one of several, together with many Cherry Laurels and Holm Oaks – both bêtes-noires to Norman and to later managers.

Holm Oaks love the mild climate, tolerate blown salt spray and have spread extensively. With their dense evergreen foliage and tough waxy leaves which decay very slowly, they suppress ground vegetation, even limiting Ivy. It is not the only exotic that has spread widely, as was emphasised by J F (Hamish) Archibald in the first Management Plan in 1965. He listed the woody species found in the Reserve and the maximum girth of some of them. Among the 89 species he found, 11 were conifers and of the larger trees, 30 were exotics and only 33 natives.

Chapter 1

I met Hamish by chance at a Tree Wardens Conference and heard of his work with the Nature Conservancy Council's woodlands unit before he came to Furzebrook, near Wareham, as Assistant Regional Officer for the South West. He recalled how Dr MacFadyen, the Council's Chief Geologist, had wanted to get rid of *"all the damned vegetation"* to reveal the geology. He also had stories about some of the exotics. He had found an unusual yellowish fruited Blackberry, had sent it to Kew, where it was identified as a species from the Himalayas. Its origins became clearer when, talking to the Allhusens, then living at Pinhay, he found that they knew the plant as Rambling Rupert after an uncle who brought back a range of plants from distant explorations. One of the most exotic plants was a Banana, indicative of the mild local climate. Major Allhusen had not been happy when, having given some scouts permission to camp, they had roofed their shelter with Banana leaves.

Hamish had notes about two lime kilns at the extreme west end of the Reserve and another, below Rousdon, at the characteristically precise map reference 29789010. The nearby pumping station, built in 1890 to provide the Peek family mansion with water, not only for direct use, but also for storage in the event of fire. The pumps were powered by coal brought down a track through Charton Goyle. Hamish had another slip of paper with a quote from the Lyme historian George Roberts in 1823 *"It is said that a farmer, who lately quitted an estate between Lyme and Seaton, received a traditional account, handed down in the family from father to son, that one of his ancestors escaped with his life when it was endangered by huge fragments of cliff which, becoming detached, rolled with immense violence into the plain and killed several head of cattle which he was driving"*.

The Rousdon Estate now own a large section of the central Undercliff, while the Allhusens own a larger part extending towards Lyme Regis. They bought Pinhay House and this extensive land in 1892 and sixty years on Major Ormsby Allhusen, one-time High Sheriff of Devon, was involved in the establishment of the National Nature Reserve. His widow, Dracaena, lived for many years in Lynch Cottage, previously home to voluntary warden Laurie Pritchard. On a lovely day in February 2001, I had arranged to meet Dracaena to hear about her love of the Undercliff and her knowledge of the land above it. A small part of this land is a steep field of unimproved calcareous grassland which is probably much as the Undercliff below had been when grazing there was extensive and before tree cover had taken over. In spring, the field is full of Cowslips and later in the year, scattered Autumn Gentian and Autumn Lady's-tresses are part of the community. A quick analysis of the plants one August showed the dominance of a range of Hawkbits among the grasses, with plenty of Rockrose and Salad Burnet together with other chalk lovers including Hoary Plantain, Eyebright, Squinancy-wort and Stemless Thistle. It is also a good site for Horseshoe Vetch which Tom Wallace recorded there back in the 1950s. The narrow ungrazed southern strip of the field, between a protective fence and the vertical cliff edge already showed the early stages of colonisation by trees, as young Ash and Holm Oak were replacing the grassland and the Rock-rose covered anthills. The trees are now a good deal larger and the flowering plants much reduced.

When I asked Dracaena about local changes since she had come to live in Devon in 1940, she thought movements below Pinhay had been the worst, as the steps going down from the pumping station to the beach kept on disappearing. There had been so much change along the track to Lyme that she could no longer find where Ames' double walls had been. The plants had not changed that much but there was *"too much Ash, too much Sycamore. The forestry had rented land out there and my father-in-law had planted a bit but not my husband. He had sold quite a few trees in the fifties, but it was difficult to get them out, so we did not get much for them"*.

The movement below Pinhay had always been the worst

She maintained that it had always been possible to walk through the Undercliff, as she had first done in 1938, but people came *"in the wrong sort of shoes"*. Incidentally, she was given new boots on her 90th birthday and had worn them out before she died. She knew Norman Barns, but not well, and *"oh yes, she knew Tom Wallace. Headmaster, Keith Moore, was another who was very keen on the Undercliff, as was John Fowles"*.

13

Chapter 1

The Crimea Seat

Among a variety of papers she showed me, were some relating to her husband's acquisition of the foreshore, an impressive set of rainfall records dating back to 1892, and one of Swete's paintings of the view from Pinhay Cliffs towards Golden Cap and the Dorset coast. She also had documents relating to Ames' purchase of Pinhay, Clevelands to him, and other papers linked to its later sale. There was a map, prepared as evidence against him, which showed the line of the obstructive walls he had built.

Dracaena then took me out along the cliff top to show me a strange building, made of unknapped flints, where Ames had sat at times during the Crimean War looking out for Russian ships which he believed would appear in Lyme Bay. From the Crimea seat, as the building is known, we too could look out over Pinhay Warren, towards Seven Rock Point and Lyme Bay. The view reminded her of the time a fossil Ichthyosaur had been found near the low tide mark. As the owner of the foreshore, she had given permission for its extraction from the Jurassic Blue Lias so long as, if well preserved, it was given to the Philpot Museum. History and prehistory are both parts of the Undercliff story with prehistory reported in lively style by the distinguished French anatomist Georges Cuvier (1769–1832) who declared, on seeing a lithograph of the Plesiosaur found by Mary Anning in 1823 *"verily this is altogether the most monstrous animal that has yet been found amid the ruins of a former world. It has a Lizard's head, a Chameleon's ribs, a Whale's paddles, while its neck was of enormous length like a Serpent tacked onto its body"*.

The lithograph had been prepared by William Conybeare, palaeontologist and Rector of Axminster, who first used the name Plesiosaur or *"nearly reptile"*. With his knowledge of fossils, he thought the weird animal to be nearer to the modern reptiles than the Ichthyosaurs were. Later, Cuvier was ready to believe anything that came from Lyme. 160 years after his description, exciting remains were still being found there, like the fine specimen found by Peter Langham, west of the town.

In his book *The Dragon Seekers* (2002) Christopher McGowan, mentioned earlier by John Fowles, describes the discovery of the largest example of *Leptonectes tenuirostris*. Peter Langham and his father, Bob, had stopped for a smoke on their way back from Pinhay Bay. Having finished his cigarette, Bob tossed the butt into the water where he saw what looked like the ribs of a fairly large Ichthyosaur which lay in a band of shale below a limestone slab. With the help of David Costin, they managed to get to the fossil, removing most of it on that first day. On return visits they recovered the remaining skeleton except for the very tip of the snout. The almost complete skeleton was 3.8 metres long and because of its arched back, it became known as 'the leaping Ichthyosaur'. Although not a new species, it had real scientific value as the largest exemplar of its type. After cleaning, a laborious and skilful business, it went on temporary display at the Philpot Museum, spent a time at Peter's private Dinosaurland, before being moved to Toronto in 1999.

The view from the Crimea Seat

High above the beach where the fine fossil was found, Ames had caused a path to be made zig zagging down past the Whitechapel Rocks. Here, dissenting Christians had met for prayers in the 1600s as Roland Brown described 200 years later. *"In those perilous times the rocks sheltered those who unable to worship God according to the dictates of conscience, when the fires of persecution were burning in the land and when the iron hand of tyranny and of oppression pressed heavily on the necks of Englishmen; many a brave soldier of the Cross repaired... in these solitudes the free will offering of their hearts"*. One 'cruel oppressor' and 'indefatigable persecutor' gave information against the Nonconformists having watched them from a neighbouring pinnacle named after him as 'Jones's Chair'!

Chapter 1

George Cumberland's view from Jones' Chair in 1820

Below these pinnacles is another area of persistent instability and a spring which was to act as a water supply when a pumping station was built by the Lyme Regis Water Company in 1930 providing 2,500m^3/day for the town. The station passed to the East Devon Water Board in 1959 but by 1985, with growing worries about the instability of the cliff, the area became a focus for studies of landslide activity (Grainger et al 1985 and 1995). The introduction to Grainger's 1985 paper points out that *"the basic cause of the large infrequent landslipping events and the smaller, continual movements, is the presence of weak rocks in the middle of the geological succession coincident with a zone of high water pressures and seepage towards the coast. The Cretaceous aquifer of Chalk and Upper Greensand overlies Triassic and Jurassic strata dominated by impermeable mud rocks. Undercutting of the base of the cliff nearby removes the debris deposited there by landslide activity"*. Finally, after landslips in the wet winter of 2000-2001 the station had to be abandoned.

From the remains of the pumping station, obscured by the invasive Buddleia, the coast path climbs towards Whitlands, or more accurately to the site of West Cliff Cottage, where, in 2005, removal of a large tree which had fallen on, and perversely stabilised, the remains of the cottage, led to the collapse of another critical wall. The 1840 Tithe Map shows the cottage, with a couple of orchards and two

West Cliff cottage about 1890

fields surrounded by 90 acres of pasture towards Pinhay and another 89 acres of the same on the cliffs to the south.

The old coast path went down towards those cliffs, as does the present route to Humble Glades. It passed through an area of acid soil derived from cliff material which fell in the 1840 Whitlands Slip. By the path, Hard Fern replaces Hart's Tongue as the dominant fern, while Silver and Downy Birch trees favour the acid conditions.

When William White reached this area on his long walk, he crossed a field with a stile but was told by a man digging that *"I could go no further along the Undercliff, the path soon disappeared, everything was left to grow just as it liked, no-one could get through the tangle, or go around it. There was Pinhay he said, and there Whitlands, pointing to the two sides of the amphitheatre... True enough the path soon disappeared, and brambles and briars had it all their own way... but it was possible to circumvent the thorny barriers 'though not without labour and rough scramblings'; exertion well recompensed by the sight of wild solitudes and rich hanging woods"*.

It was much the same in 1929 when the *Bridport and Lyme Regis News* (on 12 April) reported that *"The Council is contemplating restoring the landslip footpath between Lyme Regis and Seaton. It is a delightful path, perhaps one of the most beautiful on the south coast. The portion lying between Pinhay Bay and Whitlands is, however, in a sorry condition and its renewal would entail considerable labour and expense"*.

For some way past the area of the Whitlands Slip *"the land is very torn with large fissures, many hidden"* until a long series of steps takes the path down to another pumping station, this one built by the Peek family. They had bought the land for their estate in 1874. In 2001, after 60 years as Allhallows School, and, later, College, the land changed hands again and Rousdon Estate residents can now make their way down to the 'private' beach below the pumping station. For some time after the change of ownership the final part of the descent was down a ladder

Part of the pumping station (chimney extreme left) below open cliff in about 1920

Chapter 1

which had been the fire escape from the girls' accommodation at Allhallows.

Not surprisingly, the Peeks had a significant fear of fire as there was no regular fire service at the time, and for that and other reasons, built a reservoir from which they could get water up to the mansion. To do this they needed coal which was brought down a track from Charton. It fuelled a fire which generated the energy to pump water up the hill. The Peeks' boatman and gamekeeper, Sam Edwards, lived in the cottage until its garden subsided overnight in October 1911 leaving the cottage on the edge of a 30 foot precipice. Over 100 years later, it had disappeared down the cliff. Below and to the west a 'harbour' had been blasted out from the rock to shelter the Peeks' yacht. From here, Sam would set out to catch Lobsters.

"... leaving the cottage on the edge of a 30 foot precipice."

The Peek's "harbour".

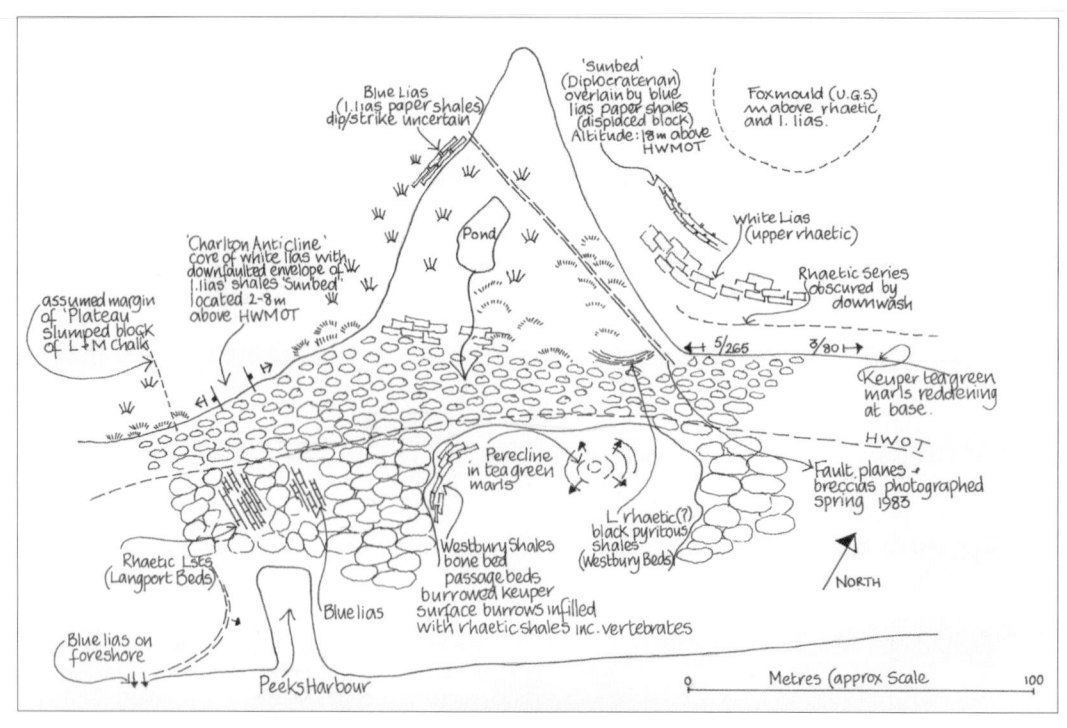

In 1988 Keith Moore, later to be Headmaster of Allhallows, sketched the harbour and its surroundings with which he was as familiar as anyone. When interviewed in 2001 he reckoned to have seen parent rock exposed, even if only briefly, all along the Undercliff. The only exceptions were Humble Point, where fallen rocks from the 1840 slip had buried the old sea cliff, and the Plateau, a great mass of chalk which had subsided in the remote past. The Chalk overlies Limestone and Mudstones, but a major fall had taken these rocks down to sea level leaving the flower-rich mass above the sea.

Keith emphasised the need for long-continued observations, a plea echoed by Norman Barns, who described how the pinnacle on Dowlands Cliff had overstepped the inland cliff by at least 15 feet but 20 years later, in August 1997, its highest point was some 12 feet below the cliff top. He told his walk leaders to continue for a mile below these cliffs until coming to an obscure path which had been a significant cart track used to bring down livestock and food for those living in Landslip Cottage and to take up Ivy and hay for sheep fodder. From Rousdon to Axmouth was all pasture until the turn of the century, Norman told his potential walk leaders. He continued by describing the now ruined cottage and a clearing extending along both sides of the path. That was in 1986 but by 2002 the Hazel needed to be coppiced again. Just beyond is the Resting Stone where women and children sat on their way to or from taking food and drink to the workers. In 1940, children were, again, involved acting as 'runners' for the Home Guard who were on the lookout for any enemy activity in the bay. Now, the sea view has long gone for, once again, trees have taken over.

Six years later there was German activity in the area when two prisoners of war escaped from a farm at Clyst Honiton on 1 December and set up camp in the Undercliff. A report said that their camp was 200 yards below Landslip Cottage but 50 years later Dracaena Allhusen, who had actually visited the German's camp, placed it ten minutes west of the cottage, an even more secluded site. It was a bitterly cold winter, with the Axe estuary frozen over, but the escaped prisoners had erected a shelter in a large crevice equipped with hay and a blanket, cooking utensils and an oil stove. Although the search for them only began on 23 February, after a burglary at Pinhay House, beach huts at Seaton had been raided two months earlier and Annie Gapper at Landslip Cottage had reported that water in their well had been disturbed making it difficult to use. Not surprisingly, Major Allhusen had useful contacts and was able to arrange for two bloodhounds,

Marcus of Bratton and his sister, Dinah, to be brought with their owner, from Bratton Clovelly, near Okehampton, in a police car.

By the 24th the police were spread out from East Lodge at Allhallows, all the way down to the shore, and if they were as described *"about ten yards apart"* there must have been more than 50 of them. The next day they were joined by 100 Marines. By then the bloodhounds had found the campsite and the remains of chickens and sheep but the prisoners had left, evidently recently, for there was still warm water on the oil stove. The next issue of *Pulman's* reported the capture of Hans Heinel, near Weymouth, but it was another fortnight before Ruprecht Reide was found on a straw bed in a farm outbuilding near Corfe Castle. He put up no resistance.

Ten years later, memories linked to the prisoners were recalled in the paper with Stan Carter of Home Farm, Rousdon, remembering that the Germans had stolen a couple of oars hoping to escape by sea and that another prisoner, working on Home Farm, had always been keen to fetch the cows in if they were grazing near the Undercliff. The article also mentioned that Annie Gapper remembered that they had kept a chicken, tied by its legs to a tree, to provide them with eggs.

An unlikely follow-up came after another ten years when Derek Stevens at Interlink Publishing in Mayfair wrote to the Ministry of Defence asking about the prisoners. The Ministry had no information as POW records had been sent to Germany. Soon, a reply from Berlin provided the addresses of the pair of escapees. No contact seems to have been made but another 30 years on Derek, now a reporter with *Pulman's*, was able to write that Herr Reide, now 82, lived in the ex-East Germany and that 75 years old, Herr Heinel, lived in Ludwigshafen: his account also mentioned *"Chickens clucking in the darkness"*. His article led Jo Draper, Curator at the Philpot Museum, to contact Mrs Allhusen who had actually been to the German's campsite. She had found no fitted furniture and made a firm denial that a side of bacon, not available on the rations of the time, had been taken from Pinhay House.

It was also near the site of Landslip Cottage that short-sighted Elaine Franks had almost fallen into a bramble filled pit. At the time, Elaine was preparing the illustrations for her *Sketchbook of the Undercliffs* (1989). The pit was actually a large sheep-wash built between 1790 and 1800. It could be filled with water from the same spring that supplied the cottage. Farmers would have brought their sheep from miles around, driven them down a ramp into a sunken bath, where they were washed and wiped down with Fuller's earth, a degreaser, and lime to make them whiter. To do the washing, shepherds would have stood with their sheep in a good depth of water and afterwards might well have needed something like the 16th century Yorkshire sheep-wash warmer; warm milk, ale, breadcrumbs, nutmeg and enough pepper to make it hot.

Soon after Elaine's potential fall, Norman Barns and Terry Sweeney worked to clear the brambles, accumulated leaves and young trees, before repairing the stonework. This meant that someone, I think it was Terry, had to wheel the bags of cement all the way, along the tortuous coast path, from below Allhallows.

Norman had been much involved in helping to shape Elaine's book, and the end of a letter from her to him suggests here willing acceptance of Norman's comments.

In autumn 1999, an early Axe Vale Conservation Society work party cleared it again and treated the Sycamore stumps to prevent regrowth. Repairs then, and since, have used traditional lime mortar and much of the surrounding area has been cleared of trees potentially opening up views to the sea and reducing the problem of the accumulating leaves. A fairly obscure path passed from the cleared area down to the fossil rich 'Slabs'. Cliff falls in 2014 destroyed the old route down and now, August 2020, the seaward end of the replacement route has collapsed.

Past the sheep-wash and the 'Resting Stone' the coast path leads to the 'Avenue', an unusually straight route. Beside it, rocks from the 1839 landslip litter the bramble covered slopes around the path, which eventually curves below the east end of Goat Island where, as well as in Culverhole and the Chasm, I counted birds as part of the national Common Bird Survey which attempted to record all the breeding birds in specific areas. Dense bramble scrub was, and is, always full of summer Blackcaps, but Nightingales, which had long been a feature of the Underclif, were only present in 1994, the first year of the study.

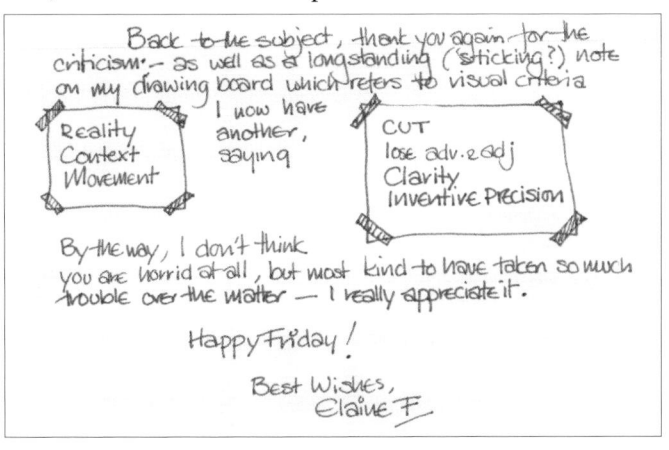

In front of Goat Island, a dense mass of scrub species compete with sun seeking climbers, scrambling between them. In 2014, steps on this section of path repeatedly collapsed and after searching for possible alternative routes, Reserve Manager, Tom Sunderland, chose to create a path over Goat Island despite some fears about damage to its plant communities. The chosen route also involved a long descent with the steps needing a lot of carrying of durable wood, but even before the new path was opened, they had efficiently been put in place. There was a delay in creating an extension to the cliff top path as the District Council had to negotiate for access across what had been Bindon Estate land.

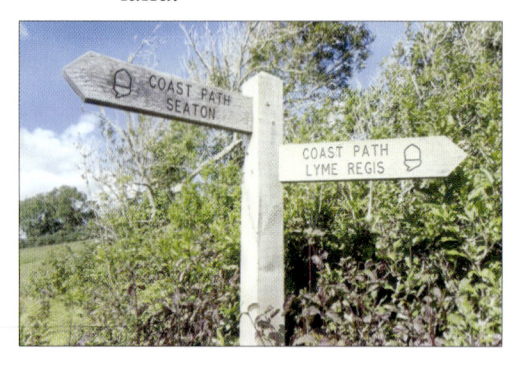

Tom chose to create a path over Goat Island and down a long descent towards Lyme Regis

Access to Culverhole was still needed for habitat maintenance and species recording so the coast path was not totally abandoned and an old, somewhat precarious route, down to the sea was kept open. Culverhole is a classic 'soft cliff' site and, as such, is highly unstable and constantly on the move. Among the plants, Marsh Fragrant Orchid and Marsh Helleborine grow with Black Bog-rush and Great Horsetail. Landslipping does much to maintain the conditions, also favoured by sun loving invertebrates, but individuals and work parties sometimes lend a hand controlling some of the scrub on the relatively stable areas.

Norman mentioned 5 miles 124 yards as the distance from the entry stile at Lyme to the emergence of the coast path onto agricultural land at Bindon. This meant that there was no access to the final overgrown mile and a half towards the Axe. In 1931 the path along there had been closed after a fall of thousands of tons on 2nd April. The fallen rock formed a 300 yard long promontory below Haven Cliff according to the report in the *Devon and Exeter Gazette*. For several months in 2001, there was no access at all along the coast path, which was closed during the foot and mouth disease outbreak, but when people returned, all had changed for some 600 metres of intact land up to three metres wide had subsided between three and five metres during the epidemic. The path continues towards Seaton, crossing the golf course, and the old concrete bridge across the Axe.

History and Exploration

John Pitts who spent ten years working on the geomorphology of the cliffs commented, in his PhD thesis, on an anomaly in the accounts of landslipping along Haven Cliff as Arber (1940) contended that it was comparatively recent, whereas Griffiths (1967) considered it to be the cause of Axmouth's decline as a port in the 12th century with the cliff path blocking the estuary and the harbour. Both could well be right for falls are frequent here. In his thesis, Pitts described events there in the wet winter of 1976 – 1977 *"Slipped material, overlying and obscuring much of the Keuper Marl, re-slipped extensively, re-exposing much of the Marl. Seepage was much in evidence with gullying on the upper surface of the Keuper rapidly developing and much standing water accumulating in the undulations of the slipped material at the cliff foot. Large earth flows were developing at the toe in parts of Haven Cliff displacing highly disturbed and wet Cretaceous material from the higher parts of the cliff".*

View from the Crimea Seat.

CHAPTER TWO
Early Days of the National Nature Reserve

In 1942, those who wanted to emphasise the scientific aspects of nature conservation included members of the British Ecological Society who set up the Nature Reserves Investigation Committee (NRIC). Its President was Cyril Diver who in the parliamentary recess put his enthusiasm into invertebrate studies and the ecology of Dorset's South Haven Peninsula. By contrast, the supporters of John Dower, who had campaigned through the thirties for the creation of National Parks and Areas of Outstanding Natural Beauty (AONB), put their emphasis on the recreational and amenity value of protected areas without detailed reference to the conservation of wildlife.

Also in 1942, the Nature Reserves Committee was established with the object of helping the Investigation Committee to select *"places suitable for preservation with information about them"*. Letters from the Chairman, Arthur Tansley, to the 360 members of the Ecological Society asked for opinions on what should be preserved and the procedures for protecting the potential conservation areas. Eventually, in 1945, the NRIC proposed 55 National Nature Reserves and 15 Habitat Reserves. The Undercliffs did not feature in either category.

The word conservation was being increasingly used in place of preservation in the belief that it would promote a positive image of science, of the widespread enjoyment of nature and the promotion of education about the natural world. The Ecological Society was also concerned about wider issues pressing the case for an authority *"to organise and direct ecological research essential for the progress of scientific knowledge and economic well-being of the country and to secure a degree of wildlife conservation which is the necessary material of such research"*.

An important leader of the conservation world for many years would be E M (Max) Nicholson (1904-2003) who had written about birds while at Oxford reading history. He had organised the first of the Heronry censuses which in turn led, with other team enquiries, to the establishment of the British Trust for Ornithology in 1933. Later, when writing for the *Weekend Review*, he responded to the politicians' call for a more positive attitude to their policies when he drafted 'A National Plan for Britain'. This led to a pioneering 'think-tank' with Nicholson as Director. Various roles in the War, mainly involving shipping, took him to important conferences, such as Yalta and Potsdam, to which he always took his binoculars.

After the War, when Labour won a convincing majority in the 1945 Election, he joined Herbert Morrison, later Deputy Prime Minister, as his principal advisor. Morrison, a Londoner with little knowledge of the natural world, had been a pioneer of the Green Belt concept in the 1930s when Leader of the London County Council. Working with Morrison, Nicholson was well placed to promote nature conservation as one of his scientific responsibilities. In April 1948, Morrison, wanting peacetime science to bear on both town and country planning and agricultural policy, regretted that *"the Government is constantly taking action liable permanently to affect the fauna, flora and even the geography of the country without having at its disposal any channel of authoritative scientific advice about the probable results, such as is available in all other fields of natural science"*.

In February 1949, Morrison announced that the Nature Conservation Board and Biological Service would be called by a more convenient title 'The Nature Conservancy' and that it would be on a similar footing, but on a smaller scale, as the Medical and Agricultural Research Councils. Four months earlier Cyril Diver had been *"dragged, very reluctantly"* from his job in the House of Commons to become Director General (Designate) of what would become The Nature Conservancy. In August, he emphasised that nature conservation was not intended to be a purely negative activity, the overriding intention was positive management on scientific lines.

The Conservancy was granted its Royal Charter on 19 May 1949 with its purpose *"to provide scientific advice on the conservation and control of the natural flora and fauna of Great Britain; to establish, maintain and manage Nature Reserves in Great Britain including the maintenance of physical features of scientific interest; and to organise and develop the research and scientific services related thereto"*.

The speed with which the Conservancy acted in sending L J Watson from the Ministry of Town and Country Planning down to Devon to investigate the proposed acquisition of the coastal area between Seaton and Lyme Regis suggests their doubts about the intentions of the Forestry Commission, which appeared to many to be over-eager to gain land for conifer plantations. Watson's note EW/M/49/14 dated 30/09/1949 recording his visit to Pinhay and Rousdon on 8 July 1949 is just over two sides of A4.

"Whether the Forestry Commission's proposed acquisition would be likely to affect scientific and/or amenity interests, seems to me a question to which it is quite impossible to give a direct answer without knowing more details of the system of forestry which is contemplated…".

"Personally, I can see no objection to the Commission acquiring the existing woodland areas provided that the intention is to maintain their existing hardwood character. The woods are in need of management and insofar as their succession would be ensured, the acquisition of the woods would be an advantage. If, on the other hand, it is not the intention to retain the hardwood character, but to replant with conifers, the position would be entirely different, since the usual timber production type of conifer planting would, I feel, detract very considerably from the present beauty of the area".

His note goes on to consider the scientific interests of the Sidmouth to White Nothe Scientific Area quoting Cmnd 7122 which stated: -

"The cliff sections of the coastline, together with that of Purbeck, show a succession of geological formations from the base of the Lias upwards, so complete, clearly visible and comparatively easy of access, that they have been, and are continually, used as a field textbook by teachers, students and amateurs alike. The cliffs, clifftop grassland and well-developed Undercliffs have a rich and varied fauna and flora."

"The rapidly changing succession of geological formations and the movements to which they have been subjected are largely responsible for the wonderful coast scenery. This coast has been described as 'the finest geological museum and training ground, for its size, in the country'. The educational value which the area affords is no doubt a factor of considerable importance."

Watson was also a painter and his portrait of 'the Late Sir Arthur Tansley FRS' is the first illustration in Sir Dudley Stamp's *Nature*

Chapter 2

Conservation in Britain. Another of Watson's paintings forms the frontispiece of Peter Marren's *The New Naturalists*. This shows Tansley and A S Watt closely examining some small specimen while E B Ford, also an FRS, watches from ground level. Cyril Diver is on his knees searching for his beloved invertebrates. In the previous year Ford's *Butterflies* had been the first, and one of the most successful titles in the *New Naturalist* series which now numbers well over a hundred titles.

Sir Arthur Tansley FRS

Three months after Watson had reported from the Undercliff, Tansley, who had a cottage nearby in Beer, visited on 16 October before describing the *"Scientific Values of the Landslip (Undercliff) between Lyme Regis and Seaton". - "These landslips, which form a continuous strip about four miles long and from 1/8th to more than a quarter of a mile in width, are the largest and most important on the coast of Great Britain. The whole area is unique of its kind: first because the character of the rocks involved, and the manner of their slipping, has led to the preservation of great masses of undisintegrated material instead of the usual mixed and often comminuted debris; and secondly because in this case alone we have historical records of the great Bindon/ Dowlands slip which occurred in 1839, including accounts from eye witnesses of the actual occurrence and detailed descriptions of the results by leading geologists of the day, together with numerous sketches and other illustrations. Nothing like the great chasm below Dowlands, created by the subsidence of some twenty acres of land separating fifteen acres of cultivated clifftop from the land behind, can be met elsewhere in Britain, and this Chasm has long been famous as a classical phenomenon following the contemporary descriptions by Conybeare,*

The Great Dowlands Slip looking south, September 1949.

28

Buckland, Lyell and others. As such, it has been constantly visited by British and foreign students for more than a century. It is, therefore, important to all geologists that this feature should be preserved, undisturbed, so that its future history may be followed. Successive landslips which have certainly been going on for many centuries, and probably millennia, form the rest of the Undercliff area, portions of which are always more or less on the move. Considerable individual slips are recorded from 1765 and as recently as 1886.

"On the eastern part of the landslip area (Pinhay and Ware) beds of the Lower Lias (Blue and White Lias), the formations over which the overlying Chalk and Upper Greensand have slipped, are well exposed in the cliffs of the actual coastline, in more than one place. It is important to preserve access to these fossiliferous exposures as Dr Haskell, the leading British authority on Jurassic palaeontology, points out for comparison with the exposures of Lias further east along the Dorset coast.

"Nearly the whole of the area is now covered with vegetation, and the slipping of different kinds of harder rocks (Greensand and Chalk) over the softer shaly Lias shales has given rise to a very uneven terrain, and extremely varied plant habitats ranging from well drained ground to local marshes and from sheltered hollows to slopes exposed to the full force of the Channel gales. The most exposed slopes nearer the sea are covered with low wind-cut scrub and the shoots of any tree or shrub rising above the general level of this scrub are pulled back. A number of dead trees are present in the most exposed places. In the areas nearer the cliffs, which have some protection, the tree growth is quite good. Here, between Pinhay and Whitlands, there is a good deal of well grown Beech, which was apparently planted about a century ago at the time Pinhay House was built, and much of it is now being felled. Elsewhere, under the cliffs and in the Dowlands Chasm, Ash has sown itself abundantly, and in several places, natural Ash woods, including quite large trees and good undergrowth, have developed. These are evidently maintaining themselves by natural regeneration, saplings and trees of all ages being present. They are, in fact, examples of self-sown virgin woodland, an extremely rare phenomenon in present day Britain, and, as such, are naturally of high interest and their preservation of great ecological importance.

"The flora of the landslip is naturally very varied: it appears to be quite typical but has never been studied in detail. The emphasis of *interest is clearly on the vegetation and its development rather than on the occurrence of rare or unusual species. There is no part of the British coast on which there exists such a great extent of varied wild vegetation almost unspoiled by human interference, and it is a rich field for future research on plant ecology and vegetation development.*

"A number of interesting birds (both resident and migrant) and lepidoptera have been recorded from the area but very little work seems to have been done on any of the animals.

"The scientific case for preservation rests primarily on the great geological, physiographical, and ecological value of this unique area. The Dowlands Slip can justifiably be described as a Geological Monument of the first importance 'though of a different kind from most because it is a record of a physical event, not a specimen of a particular rock or rock-form.

"It is relevant to call attention to the instability of the area. As Arber (1940) writes "no part of the Axmouth landslip can be considered secure ground". Heavy and long continued rains, such as preceded the 1765 and 1839 slips, may at any time bring about further considerable falls from the cliffs or movement outwards of great blocks of land. All the conditions for such slips are permanent here because they depend on the geological structure of the region. Minor slips, followed by erosion, constantly occur now on the steeper slopes, and widening cracks are widespread in the turf. If areas are cleared for tree felling, erosion will certainly occur on the slopes, whose surface will quickly become unstable. Increased percolation will also encourage slips of increasing magnitude.

"In a word, the area is very 'dangerous', for exploitation but ideally suited for research."

Tansley together with Cyril Diver, Sir William Taylor and Mr Sale all came to the Landslip on 25 July 1960. They admired the view across the Chasm. Because of the unstable nature of the ground, and scattered arrangement of the sheltered patches of land, Sir William thought the area unsuitable for forestry.

In the afternoon, a larger party including Major Allhusen, Chairman of the East Devon Planning Committee and owner of the Pinhay Estate, went to the beach near Humble Point and were shown a dense mass of scrub which the Major thought should be cleared and planted. There was no doubt that below Pinhay good timber in the form of a high

Beech plantation could be produced in a sheltered part of the slip. Tansley's summary of the scientific value of the Landslip had been circulated and it was agreed that the stretch of coast fully merited the status of a National Nature Reserve. It would have maximum value if it extended along the whole length of the slipped face, but the western half contained examples of most of the features of prime interest.

For the Conservancy, the important question remained; should the whole length of the Undercliff be included in the Reserve? B T Ward who had visited in June 1953 had no doubt. *"The Dowlands and Pinhay Landslips shall be conserved as a natural habitat for the features of vegetation which are here preserved and can maintain themselves and that the Ware Coast, west of Underhill Farm, be treated as one with the Dowlands and Pinhay Landslips.*

"The Undercliff is extremely irregular with ravines running parallel to the cliff line, intersected by small abrupt valleys, some containing streams. The largest of these is that which descends a little to the east of Rousdon. In the hollows of the ravines, small pools or marshes occur. These appear to be frequently altered by the settlement of the land. There are also numerous wet flushes which provide suitable damp habitats for many plants especially the Bog Pimpernel and Marsh Helleborine. The Undercliff varies greatly in width with an average of about a quarter of a mile. In the wider parts it is usually less steep, but much wetter, and the marshes and pools become more abundant. Because of the irregular nature of the terrain and the dense growth of the shrubs it is frequently impossible to leave the confines of the path.

Above - Bog Pimpernel
Below - Marsh Helleborine

"Towards the western end of the area described on the map as 'Landslip Cliffs', just above Culverhole Point, there are some more open stretches covered with large boulders with deep chasms between. There is also a more open area on the slope arising from the beach of Charton Bay. On these open places the Evening Primrose is established and also Pampas Grass which appears to be spreading. Just on the

extreme western extremity on an open grassy place Corky-fruited Water Dropwort occurs in limited quantity.

"On some of the open spaces, species of orchid were observed but it was not possible to approach them sufficiently closely to be sure of the identity. Nevertheless, the following species were noted: Marsh Helleborine, Bee Orchid, Early Purple Orchid, Pyramidal Orchid and Fragrant Orchid. In several places slipping has resulted in claggy hollows encouraging Sallows with Bulrush, Spiked Rush, Juncus and Black Bog-Rush with Yellow-wort and abundant Viper's-Bugloss."

Later in 1953, Norman Moore was appointed the Nature Conservancy's Regional Officer for the South West, an area stretching from Hereford and Gloucester to the Scilly Isles. On 7 October, he found time for a first visit to the Undercliff, after trying his best to find out something about its natural history from any potential source. Neither Lyme Regis nor Exeter Museums knew anything about it and nor did the Zoology Department of the University College of the South West. The Curator of Exeter Museum was asked to look for any information about the area but found that apparently no biological studies had been made there.

Pyramidal Orchid

Norman Moore's walk along the coast path from Lyme was not productive but a Little Owl would be a nice surprise today and the presence of four Stonechat suggests that the path then was much less enclosed. When able to look along the beach, or out to sea, he saw nothing exciting. The most frequent butterflies were Commas while hundreds of Silver Y moths suggested a recent influx. Among the plants there were many exotics. It was hardly an exciting day, but he was hopeful that with such a wealth of microhabitats, rarities would be found. He also noted that Landslip Cottage was almost a ruin, unlocked and with plaster falling from the ceilings and with walls crumbling.

Stonechat

Norman's next move was to contact the Rev. F C Butters who had earlier been Secretary of the Devon Bird Watching Society. Butters' reply was not encouraging as little was known and *"As for matters botanical, I am not entitled to speak"*. He knew of no Peregrine eyries but thought that Mr E H Ware, his successor as Secretary of Devon Birds, should be approached and that Sir Arthur Tansley might be useful.

Next year, Tom Wallace, soon to join the staff at Allhallows School, led a large party of Devon naturalists into the Undercliff on 26 June. Keble Martin was back while Professor L A Harvey, whose *New Naturalist* book on Dartmoor had been published that year, Malcolm Spooner, leading authority on Hymenoptera, the Reverend R F (Diatoms) Bastow and Mrs Bolitho were among the 65 walkers. Starting below Dowlands Farm they found plants in bare chalky places near the clifftop that would not be there today. Not only did they find the Early Gentian, again, but also many typical chalk lovers. As they were mainly in tetrads SY 28.89 and 29.89, they may well have found their plants on the Plateau.

Pioneering ornithologist Max Nicholson was a key figure in the founding of the BTO in 1933. This painting hangs in the reception area of the Trust's HQ.

Although Tom would later make enormous contributions to knowledge of the Undercliff, these would be dwarfed by the national achievements of Max Nicholson. Describing the early days of conservation, writers in *Nature's Conscience* (2015) focused on Arthur Tansley and *"the small Cockney-spoken, intelligent, innovative, organisationally supremely efficient and highly energetic Edward Max Nicholson"*. Some of Nicholson's early achievements before he became Director General of the Nature Conservancy in 1951 have been mentioned. In that year his *Birds and Men* (*New Naturalist 17*) was published and ten years later he was involved in founding WWF, The World Wide Fund for Nature. He went on to organise the 'Countryside in 1970' conferences, to edit *British Birds* and part edit the nine-volume *Birds of the Western Palearctic (BWP)* produced between 1977 and 1994. In his eighties he continued to champion urban ecology, particularly in London. On his 90th birthday the *"stalwart founder member of the team"* was presented with first copies of Volumes VIII and IX of *BWP* by the Editing Board of the new (1998) Concise Edition of the major work.

Nicholson visited the Undercliff on 17 September 1954, starting on the beach east of the mouth of the Axe. He thought the beach was rather heavy going. Later, this time by an easy walk, his party were able to reach a point looking out east along the Chasm and over the top of the Ash wood. This, he thought, was almost certainly the most interesting

33

and spectacular view in the whole Reserve and it was unfortunate that the Bindon farmer had so far resisted efforts to include provision for access to it, even by representatives of the Conservancy.

After tea at Pinhay, the group went down with Major Allhusen to his pumping station and to Pinhay Bay. The footpath from Lyme to Axmouth at this point was not at all in a good state. The Major expressed himself satisfied with the revised agreement and hoped that the Conservancy would not unduly mind members of the public wandering off the path. He thought advantage might be taken of the Long-Distance Footpath Scheme to secure financial aid for the improvement of the path which was the first section of that approach to the South Devon coast. He felt that a good footpath was the best assurance against trespassing but doubted that Devon County Council, quite definitely responsible for its maintenance, would do much without encouragement and aid. He promised to do his best to speed up the completion of the Nature Reserve Agreement.

Most of the Undercliff was finally designated as a National Nature Reserve on 16 March 1955 with the Rousdon and Charton Cliffs being added on 16 July 1956. Soon afterwards, Tom Wallace began teaching at Allhallows where he worked with the students collecting data about the wildlife of the extensive school estate which extended far into the Undercliff. Later, he produced booklets explaining something of the geology and describing the wide range of wildlife. His first report in 1963 acknowledged help from M C F Proctor of the University of Exeter with records of mosses and liverworts and from P D Orton, a well respected amateur mycologist. In 1960 Orton had been one of the leading mycologists who had drawn up *The New Checklist of British Agarics and Boleti*. Eventually Wallace had accumulated records of some 1,600 species from 21 taxonomic groups.

Amanita muscaria and birch roots

Remarkably, Orton found five fungal species allegedly new to science – *Crepidotus subtilis*, *Leriota brunneocingulata*, *Mycena tortuosa*, *Pleurotellus patelloides* and *Pluteus xanthophaeus*. Wallace found the Undercliff list of fungi particularly interesting in its richness in the genus *Mycena* and its paucity in genera, such as

Russula, Lactarius, Cortinarius and *Boletus,* the many species of which are well known associates of conifers. The distinctive Agaric, *Amanita muscaria* favours acid soils and is therefore rare in the Undercliff. As shown (previous page), it forms a symbiotic relationship with Birch roots. He reckoned that as toadstools and other larger fungi produce their fruiting bodies sporadically, and as these only last for a short time before decaying, it generally takes many years to complete a fairly full list of the species occurring in any particular locality. Ash woods, unlike those of Oak, Beech, Pine, or other conifers, do not usually contain a very rich flora of the larger fungi. Peter Orton would later co-operate on an extensive fungal survey of Tom's home patch in Membury.

When the second Management Plan appeared in 1992, Tom was offended that there was no reference to his records of species while teaching at Allhallows, or to Hamish Archibald's original 1965 Plan. Feeling undervalued, he ventured to mention evidence of his *"enthusiasm, experience and competence"* shown from 1930 when he had reported the first Black Redstarts breeding in Britain and 1933 when he joined the British Trust for Ornithology on the day of its formation. Much later, as Secretary of the Botanical Section of the Devonshire Association, he was deeply involved in the 1952 *Atlas of the British Flora* and was a founder of what is now the Devon Wildlife Trust. Tom was also on the Exmoor National Park

Allhallows, school and college for 60 years

Committee for 13 years and one of the five principal recorders for the 1984 *Atlas of the Devon Flora*.

He did not mention these activities when I talked with him 45 years after his starting at Allhallows. He was at his home in Membury and 88 at the time, but his memory was fine as he recalled his delight when Norman Moore had remarked that the Undercliff's wildlife was possibly better and more fully recorded than that of any Reserve in the South of England.

He started by recalling a visit to Victoria Street where the newly established Nature Conservancy had their headquarters. He found that they, like others, had no record of Undercliff wildlife but he suggested, knowing something of its plant life, that it should be considered as a biological, as well as geological, reserve. His important link with Allhallows and, therefore, with the Undercliff came about after a cricket match at the school. He had been 'dragged!' into the game between the Headmaster's Eleven and the Devon Dumplings. Sir Peter Watkin-Williams, who had been a chief judge in three African countries, the sort of person Tom liked, had gathered the Headmaster's team together, and on the same day Tom was persuaded to set up a Biology Department at the school.

Once there, his relations with students seem to have been both informal and productive for they were soon working with him to produce lists of species for the Nature Conservancy. *"It was ideal, when it was a nice day, we could go out and throw quadrats around. I had students with special interests in birds, one was keen on Badgers, and another on Roe Deer. Not many went for plants though; men do not seem to go for plants. Nobody bothered with toadstools, they kicked them. We did manage to poison a boy. On a foray around the grounds, he happened to fancy a nice red one, Russula emetica; he was sick for three days. They worried about it. It was all done without my knowing. He thought he would try it!*

"Ready for kicking" - *Macrolepiota mastoidea* (Parasol Mushroom)

"I kept records of all sorts on a card index system keeping up to date in that way. Experts, you see, they volunteered. The British Bryological Society, the British Ecological Society and the British

Association for the Advancement of Science all wanted to visit the place and be led". I mentioned that I had a list of the shore fauna prepared by P D Armitage. "That's right" said Tom, *"one of my students contacted L A Harvey, expert on shore life, Professor of Zoology at Exeter. I was there for a time as Keeper of the Plants. I also spent a day or two with Malcolm Spooner talking about all sorts of things but not much about Hymenoptera."*

When I mentioned D P Merrett and spiders, Tom remembered that that was University College, London who came on their post graduate conservation course. *"An awfully nice crew they were, staff came and reconnoitred for a week or two. Students produced quite good lists, but records needed vetting. They were interested in how old some of the Ash trees were; you bore a hole and count the rings... This will not do"* I said, *"you can't just start making holes, you must fill them. They agreed. A couple were sent to Axminster, bought up all the chewing gum there, went chew, chew, ger chew, and filled the holes with gum."*

M C F Proctor, author of New Naturalist books on *Pollination*, 1973 and 1996, and more recently on the *British Islands and their Vegetation* (2013) was, as we have seen, also associated with the lists. *"Dr Proctor, I shared a room with him at Exeter for two or three years, we did not always agree about things but still. Very good man on plants, absolutely, but did not know one fungus from another; low form of life, he regarded them."*

'A Low form of Life' - nobody bothered with toadstools.
Geastrum triflex.

Tom Wallace strides out.

I then claimed that with all these different sources, his own field work, students, societies and visiting experts, he kept the NCC updated and eventually produced his booklets. *"No"* he said, *"that was the trouble."*

Chapter 2

I kept a record showing the progress of the lists which were confidential because of the number of interesting records. There are people, absolute swine, you know, who have no conscience. One party I took down there, botanical section, we had a formidable Mrs Bolitho, botanist in Axminster, came upon a nice patch of Bee Orchids. All stood around admiring and talking about pollination. From among the feet came a slim hand from behind and picked off one of them. Mrs Bolitho blew her top, got hold of this girl and gave her the worst wigging, worse than any Sergeant Major.

"John Fowles was another who supplied a few bird records. Not many but he was always very nice to me. I will not have a bad word said about him. I had trouble parking and getting about, but he let me park at Underhill Farm. He dropped me little notes "the big cat has been seen again" and that sort of thing. He reckoned a leopard or so, but it was a feral cat, of course. Special memories and special days, just happy ones with children on the beach."

Tom died not long after our interview and his obituary in the old Honitonian News (Allhallows had moved from Honiton in 1936) described one aspect of teacher Tom. He sometimes arrived a few minutes late for double biology and the class pretended to work and engaged in adolescent ribaldry. When his van rumbled into the quadrangle, gossip subsided. *"Good morning, Sir"* they murmured. *"He ignored them all and spent several minutes drawing on the blackboard two huge elaborate flowers; botanically complete, in the most extravagant combination of colours his chalk box would permit. Everyone was transfixed. What on earth was this nutcase of a school master up to? Twenty two unmotivated fifth formers watched with mounting incredulity. He raised his chalk to eye lid, his eyes glanced at us from behind his horn-rimmed glasses and began to buzz like a Bumblebee. Louder and louder, he buzzed, he buzzed his way around the entire class; he buzzed the nose of every boy: bzzz! Hmmm! Bzzz! Rising to a crescendo he strode toward the right-hand flower. The buzzing stopped. Off he went again along the row of open-mouthed pupils, then he zoomed to the second flower.*

Nonchalantly he tossed the chalk back into the box. His whole demeanour changed. His eyes met ours for the first time. They shone with a warm blend of mischief and eccentricity. "This morning we are going to do pollination". His voice rang out authoritative and clear. Wrong Tom. We had just done pollination. In less than a quarter of an hour, without a word being spoken, or sentence written, 22 languorous youths of questionable intelligence, all struggling with 'O' Level indifference, had learned pollination for life."

Late in 2021 an extraordinary coincidence led to me hearing more about Tom Wallace and his students. Tom Turner wrote from America to my niece Alex, a butterfly enthusiast, about Marbled Whites, mentioning that he had developed his enthusiasm while a student at Allhallows. Knowing my interest in the area Alex sent him my details and he soon wrote to me describing some of his Undercliff experiences. I then sent him a few apposite pages from my book and he promptly wrote back with more details.

Marbled White

My colleague for some of the "finds" was my good friend John Llewellyn-Jones. We too were "poisoned" eating mushrooms (I had mentioned a sickness-inducing toadstool incident) but we knew these well and certainly would not have thrown a Russula into the frying pan. We cooked up Wood Blewits. For some reason these turned me off eating mushrooms for many years. We also discovered a rare gentian on a cliff top west of the school. The only place we found Bee Orchids was to the east along the main Undercliff Trail. (That spot would now be dense woodland.)

When Wallace found out about Turner's interest in Lepidoptera he, or the school, purchased a Robinson moth trap which Turner then ran for many months. *Tom Wallace must have kept my nightly records (see page 36) because I certainly didn't keep copies*. Tom Turner also recalled Wallace's frequent late arrival for school particularly the occasion when his Land Rover just about avoided the milkman but hit bank of a narrow Devon lane. He arrived at school *with a concussion on his forehead*.

Tom had entrusted copies of his extensive records to a local botanist with long experience of the Undercliff. This was David Allen who had grown up in Stockland and spent more than 20 years in agricultural research, principally as a plant pathologist in Africa. Having returned to his Devon roots, I had recently met him, found plenty of shared interests and now briefly talked with him about Tom and the Undercliff.

Chapter 2

He had known Tom for more than 50 years. *"In 1943 or 1944 my cousin sat on Tom's knee at some local gymkhana, and he has been a family friend ever since. He is something of a mentor to me as a biologist. I have fond memories of early forays with him; at Studland in 1958, I had to hold on to his ankles as he went head first into a pond after something or other. He was then highly mobile but later did not go out so much because of heart problems. He was a heavy pipe smoker and now at 88, he huffs and puffs a good deal, but his head is as mobile as ever". He was an extraordinarily good all-round biologist and his devotion to the Undercliff seems second to none, perhaps equalled by Norman Barns."* **David continues** *"I've been going into the Undercliff all my life with personal records from Culverhole going back to the mid-1950s. As a butterfly collector, I used to chase after Clouded Yellows and Painted Ladies above Haven Cliff but would not be able to quantify any changes in population and never saw specialities like the Chalk-hill Blues which have been recorded recently. With plant life I could quantify; at Culverhole things have held on pretty well with Marsh Fragrant Orchids and Marsh Helleborines as abundant in the 1990s as in the 1950s even if in slightly different places. It was exciting to find two or three Bee Orchids down there, and it still is, compared with hundreds at times on Goat Island. I did not find my way there until twenty years ago and have only known the Plateau for five years or so.*

Clouded Yellow

"I have not done much mycological work there, just enough to notice differences from Stockland because of alkaline soils and Ash woods. I have found nothing breathtaking but my knowledge is less than with higher plants. Any interesting things have been checked with Jeff Benn, the county recorder, and it would be good to get him to have a look himself. David also suggested lichenologist, Barbara Benfield."

Soon, both Jeff and Barbara would be searching the Reserve for their specialties, and I would be talking to Keith Moore who had been deeply involved with Allhallows and the Undercliff as geologist, teacher and Headmaster, now he would teach me a little of the local coastal geology.

"From an Original Sketch taken on the spot by a Gentleman" 1840.

CHAPTER THREE
The Great Landslip

Many descriptions of the Great Landslip relate to events in or around the Chasm but one, in the *Bath Journal* of 20 January 1840, tells of events further south and west when it relates the experience of a coastguard who had been on duty on the beach when *"he observed the sea to be in an extraordinary state of agitation. The beach on which he stood rose and fell. Amidst the breakers near the shore something vast and dark seemed to be rising from the bottom of the sea… He fled into the cliffs above. These were also trembling around him, but he joined the firmer ground, almost dead with terror. In the morning, immediately in front of the Undercliff, which though much rent and shaken, still retained its former position, there appeared a stupendous ridge of broken strata of Blue Lias, together with rocks of immense size, unmoveable by human power, covered with seaweed, shellfish and other marine productions. The elevation of this monstrous reef, extending more than a mile in length and in some places 200 yards in breadth, is not less than 40 feet from the level of the sea!"*

Another account, written by the Reverend William Conybeare, originally for Woolmer's *Exeter and Plymouth Gazette* dated 4 January 1840, but reprinted for a wider audience in the *Edinburgh Philosophical Journal*, began *"The recent season of Christmas has been marked on the neighbouring line of coast by a convulsion so remarkable from the extent, magnitude and picturesque changes it has produced in the surface and general configuration of a line of country extending at least a mile in breadth… that I cannot conceive some account of it, cannot fail to be acceptable. Although this convulsion can only be ascribed to the less dignified agency of the land springs constantly undermining the substrata yet in grandeur of the disturbances it has occasioned, it far exceeds the ravages of the earthquakes of*

Chapter 3

Calabria and almost rivals the vast volcanic fissures of the Val del Bove on the flanks of Aetna."

The account continued *"Through the course of the following day (Christmas) a great subsidence took place through the fields ranging above Bindon Undercliff, forming a deep chasm, or rather ravine, extending nearly three quarters of a mile in length, with a depth of 100-150 feet and a breadth exceeding 80 yards. Between this and the former face of the Undercliff extends a long strip exhibiting fragments of Turnip fields and separated from the track to which they once belonged by the deep intervening gulf, of which the bottom is constituted by fragments of the original surface thereon, together by the wildest confusion of inclined terraces and columnar masses, intersected by deep fissures so as to render the ground nearly impassable. The insulated strip of fields which has been mentioned is greatly rent and shattered.... The whole surface is corrugated by new ridges and furrows and traversed at every step by new fissures, and the whole line of sea cliff has completely changed the features it possessed a week ago".*

Although Conybeare was the first to write an account of the event, William Buckland, Professor of Geology at Oxford, and his wife, Mary, who were staying with Conybeare, had also been quickly on the scene and within days, Mary had completed her first painting which showed a view of the Landslip looking west to the Sidmouth hills with Goat Island made up of more than 40 separated pillars which would later coalesce while still showing lines of weakness. In the same way, Conybeare's sectional drawing showed separate blocks of chalk and cherty sandstone forming Goat Island and the pinnacles in front of it, as well as the tilted blocks in the Chasm.

Mary's illustration was one of the ten used in the historic account, both of the Bindon Landslip and of the subsequent 1840 event at Whitlands. Another was the map of Goat Island, the reef and the lagoon, prepared by surveyor William Dawson who was measuring and mapping by mid-January. In the Buckland archive in Oxford his map is annotated in pencil *"Ground plan and section of the Great Landslip measured from 15-23 January 1840 by William Dawson. Belonging to Professor Buckland 100cm x 68cm"*. Until 1990 there was some doubt whether the model in the Philpot Museum had actually been made by Dawson, but Hugh Torrens found the proof as mentioned in a letter from Muriel Arber to Liz Anne Bawden at the Museum.

The Great Landslip

The Whitlands slip, Feb 3rd 1840 (Hawkins above, a painting in Lyme Regis Museum below.)

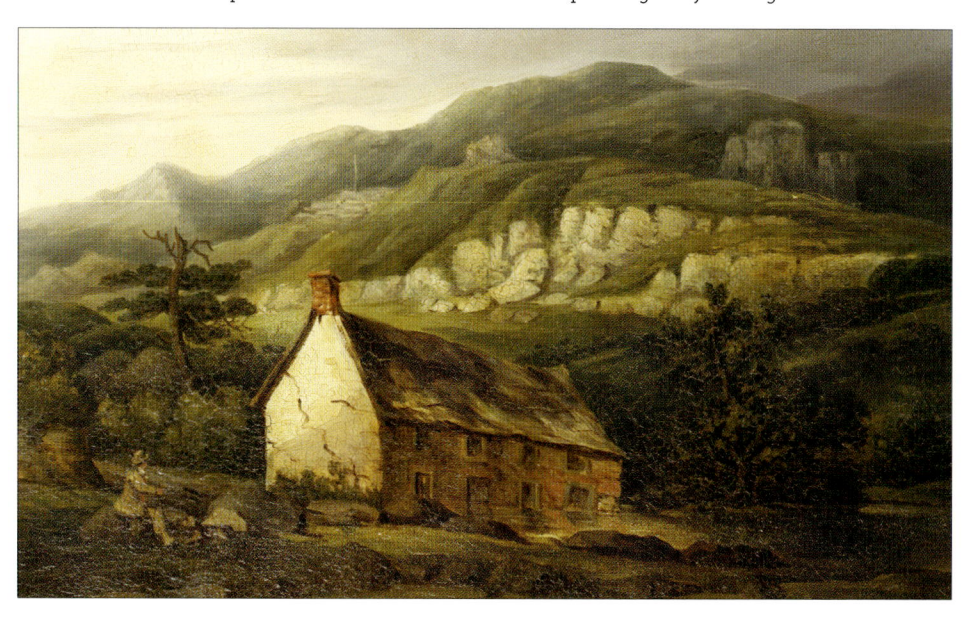

45

Chapter 3

Vast Towers and Pinnacles, view looking east, Jan 14th 1840 (Hutchinson)

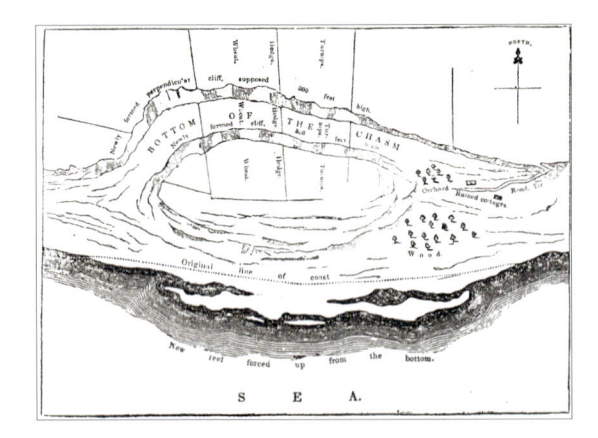

Map in the Penny Magazine, Feb 15th 1840

William Buckland lecturing at Oxford University.

"Hugh Torrens found these letters from William Dawson to Buckland in the Buckland archive in the Oxford University Museum. Hugh lent me photographs of them, and I photocopied them for myself, and have since photocopied my photocopies for John Fowles and for you. The special interest for our Museum is that in the letters of 13 February and 19 April, Dawson speaks of the model he is planning (February) and making (April) of the 1839 Landslip at Dowlands and Bindon. This is the model which is now in our Museum, on loan from Exeter, and those letters are the final proof that he actually made it himself!"

A month after Conybeare's account, Sidmouth antiquarian and artist Peter Orlando Hutchinson produced an illustrated record for the *Saturday Magazine* for 8 February. "The most notable objects which arrest the eye in the midst of the wildest part of the wilderness are vast towers and pinnacles of Chalk, which jut up from the bottom of the Chasm ... assuming the most picturesque and fantastic forms imaginable. There is another feature which is no less remarkable; the rising of a reef of rocks from the bottom of the sea, and the formation of a harbour or enclosed bay, which ranges irregularly all along the coast for a mile. When the reef first showed itself, it stood forty feet above the water, but it has been gradually sinking and is now not more than half that height."

The Penny Magazine of the Society for the Diffusion of Useful Knowledge published on 15 February looked to geology to explain the events which deserved "the consideration and interest not only of geologists, but of every thinking person. The Foxmould, as it is provincially termed in the neighbourhood, is of a loose spongy nature and imbibes all the moisture which falls on the surface in the form of rain and which filters through the upper strata from distant springs. This moisture could not subside to the Lias below owing to the imperviousness of the clay, which resists it there, and on which the water rests, as on the bed of a well puddled canal. Where the strata, therefore, show their besetting edges, or where they crop up along the declivity of the cliff, springs carry with them, slowly but inevitably, great quantities of the friable and loose earth, undermining the super incumbent strata and preparing them for a subsidence as soon as an extraordinary wet season, such as we have just experienced, shall both hurry away more of the remaining support from beneath and saturate with a greater weight of moisture the several soils above".

The article then describes how "*large areas of four different fields, with their surfaces unbroken and only somewhat thrown from their original land, still bear their crops of turnips and young wheat. The hedges too, which divided these fields, and which run to the brink of the precipice and there stop, can be traced across the portions which have subsided, and their corresponding hedges further traced across the land on the other side of the gulf".*

Other accounts and explanations of the Landslip were not slow to appear in newspapers, magazines and tracts. They ranged from the mundane to the extraordinary with a Mr Hallett claiming that *"surely it is quite as probable that the Rabbits and Foxes, which have been burrowing these for centuries, have at last brought their houses down about their ears. You know they abound in these cliffs."*

The anonymous writer of *A Brief Account of the Earthquake* was among those who saw the hand of God involved describing *"the remarkable visitation of the most high God of this land, by an earthquake, has taken place but a short distance from this part. Geologists attempt, I perceive, to account for it by natural causes for professing themselves wise have become fools (Romans I) accounting for all from nature to the entire leaving out of the God of nature who works above all natural causes, at times showing himself that he is God whose name alone is Jehovah the Most High over all the earth (Psalm IXXXIII VS 18) the first cause and the last, of all things visible and invisible".*

Similar sentiments were expressed by the writer of *Poetical Remarks*, on hearing of the Great Landslip. Near the end of 200 lines of verse, he laments that:

*"Old men will tell thee this thy God hath done,
But men of modern times thy theme to shun,
Will trace to chance or Nature's common laws
Each dire effect, indiff'rent to their cause..."*

while hoping that his verse would *"lead the reflecting mind to the Great Source of all things and among yawning cracks and detached masses would trace the Mighty Hand which will one day shake the universe and from whose presence heaven and earth will pass away like a scroll".*

A paragraph in the *Taunton Courier* which referred to flashing lights and an unbelievable stench caused some to believe a volcano to be the cause. If neither earthquake nor volcano were involved, some, as 'Sidmouthiensis' describes, would be disappointed. *"There are those who appear against all argument, to desire to delude themselves* with the idea of a 'real earthquake' and actually be proud that Old England is able to produce phenomena as wondrous as other countries can do".

William Buckland had also mentioned an earthquake when he wrote to Conybeare. "When it was confidentially affirmed on the day I was with you at Dowlands, that an earthquake had been felt at two in the morning, just at the time the great Chasm was first formed. I admitted… that supposing the state of the cliff in a condition 'apt to move' from the undermining agency of land springs and sheets of water in the Foxmould reducing it to a quicksand… the shock of an earthquake might be the cause of the movement in question. But there seems to be not the slightest proof of an earthquake at all.

"After last week's repetition of the phenomena at Whitlands, it seems worse than juvenile to talk of an earthquake where no such cause is wanted and where its interference with causes themselves fully adequate to produce all the effects that have taken place superfluous. I would not insult the majesty of an earthquake by attributing to it such a fiddly slip-slop piece of work as the undermining and destruction of an Undercliff."

Attracted by the sensational reports and the appearance of the first water colours and lithographs, visitors were soon coming to the landslip. The clifftop landowners who had lost parts of their fields also suffered as the visitors trampled their remaining crops. Learning quickly, they were soon charging six pence a visit and with crowds like "herds of sheep" they did very well.

By May, the *Dorset County Chronicle* could report that *"the famous landslip continued to prove more and more attractive. Her Majesty the Queen had been graciously pleased to honour our townsman, Mr Dunster, stationer, with an order for a series of beautiful lithographic prints depictive of the scene. The numerous and highly distinguished persons who are continually passing to and fro to explore the delightful and romantic ruins, give to our pleasant little town an appearance of gaiety and bustle which would well become a watering place of much greater extent"*. There followed a list of some of the Lords, Ladies and Honourable Gentlemen who had visited. If some of the local landowners had been slow to appreciate the potential to make money from visitors, James Chappell of Bindon Farm had got himself well organised by August when he advertised a grand reaping of the fallen crop:

Chapter 3

GRAND HOLYDAY
AT THE
LAND-SLIP,
NEAR THE
VILLAGE OF AXMOUTH.

Mr. JAMES CHAPPELL, of BINDON FARM, respectfully informs the Nobility, Gentry, and the Public generally, that agreeable to the numerous requests of the many eminent Persons and the neighbouring Gentry who have lately visited the Land-slip, he has resolved to erect Booths, & provide all suitable Refreshments for the accommodation of the Public, on *Tuesday*, the 25th ins., along the confines of the grand Chasm of the Slip, on which day the crop of Wheat growing in its bottom, (which must have sunk at the time of the eruption full **200** feet from its original situation,) is to be reaped by the Visiters, and sold on the spot in handfuls for a moderate consideration.

AFTER THE REAP,
A VARIETY OF
DIVERSIONS

will take place in a Field adjoining, for the amusement of the Visiters.

☞ *The nearest way by* **2** *miles, to the Land-slip, is to Axmouth Turnpike, from thence to Axmouth Village, and through Bindon Court.*

Dated August 14th, 1840.

Wills, Printer, Axminster.

William Dawson described the proceedings of the Grand Holiday in a letter to Professor Buckland. *"It was a really beautiful sight – the day warm and bright – and I should think a full six thousand spectators. They got up a procession which was, in my humble opinion, not quite in good taste – a committee with blue ribands around the neck – six lady reapers in white kid gloves and wreaths of artificial flowers, with sickles tied with blue and six gentlemen to match in blue vests and white trousers. They had, however, a good band of music, the effect as they wound down the zig-zag path into the valley of the chasm with the banners and the assembled thousands, lining the cliffs on both sides, was picturesque and fine. Sir W Pole was there and furnished a battery of four guns from Shute. I heard of no accident whatever and all looked pleased and happy – the young ladies reaping, however, was a failure – with the first strike of the sickle one of them cut her hand and they were so crowded upon that they soon gave over and the corn was subsequently reaped by the labourers. I met Mr Conybeare and some of his family, but they did not arrive until the reaping was over! The lady reapers symbolised nymphs of the corn-goddess – Ceres".*

Apart from visitors paying at the clifftop, people came by boat from Weymouth and Torquay, and five days later the royal steamer visited on a day that would end in tragedy. Under the heading 'The Late Melancholy Accident' the *Mercury* reported on the inquest at Axmouth investigating the death of Mr De Brue *"a fine young man and wealthy merchant of Holland"*. One witness recorded that *"I heard that the royal steamer was standing by in the bay with her Majesty and Prince Albert on board. To see her Majesty, among other persons, was the deceased gentleman. He was in a field called Bindon, part of Dowlands Farm. His horse suddenly made off and continued running to the very edge of the cliff and despite a call of 'for God's sake hold hard or you will be over the cliff' that is where he went, as he leapt from his horse and tumbled 200 feet into a quantity of underwood. When he was reached, he was alive but insensible, his eyes were closed, and he was not bleeding much. Despite being brought brandy and water and receiving attention from a medical gentleman from Lyme, he soon died"*.

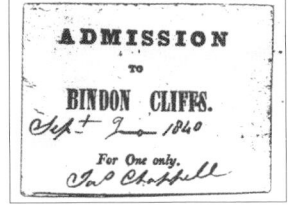

We do not know what happened to the horse but two years later another one, a handsome cart horse, suffered badly while climbing 'fearful hills' as it brought a party of four from Colyton to the Landslip.

Chapter 3

A Dinner Party at Bindon Landslip, July 17th 1842. Four top-hatted reverends and six members of the Foot family are among those at the dinner. Louisa (standing, extreme right) is one of the daughters of the Rev. Lundy Foot (at right end of the table).

The Great Landslip

The Foots were later, in 1862, to receive three sketchbooks, mainly of Dorset landscapes, from the Rev. F. Cunningham, the central top hat with his back towards the artist - who is believed to be a Mr W Porcher from East Stoke.

Chapter 3

This account of the visit comes from *The Narrative of a Tour in the Splendid Summer of 1842*. The writer, a merchant from Peckham in London, tells of a 13 week holiday visiting 13 counties. In 2000, 12 pages of *The Narrative* were sent to Jo Draper, then Curator of the Philpot Museum, by a later Mr Barnes from Hayle in Cornwall who gave me permission to use the material which now, 20 years later, I do.

"We descended the Chasm by circuitous and winding paths and if it appeared awful from the land, from the base, it was yet more appalling for in many places it overhung fearfully and wide yawning chunks of several foot width, extending nearly from the bottom to top seemed to threaten momentary destruction to those below... We went on tracing several tracks of former visitors until in some instances we were convinced of our temerity and really alarmed at our dangerous position.

"I have already said that it was an exceedingly hot day and from the great exertion and excitement we suffered greatly, so much so that, for myself, I had a most strong impression on my mind that a serious accident would be the consequence. We gained the tent of our guide... which was provided with a cask of beer. Not having a spirit licence, we could not get what would have been fitting for us, a glass of cold brandy and water, and was thus necessitated as it were to drink copiously of this bad beer which we should all have been much better without. We did not get back to our inn until five o'clock where we found a loin of mutton hot and ready, but we were all too much fatigued and heated to feel inclined to eat and, therefore, ordered our tea to be brought on with the joint and half a pint of brandy, with this we seemed a good deal refreshed, but we were all completely knocked up".

A sketch of the Bindon landslip, probably by W. Porcher, from the book given to the Foot family and dated 1843.

A year later, a sketch book provides a very different record of the Landslip. The book was presented to the daughters of the Reverend Lundy Foot of Little Bredy in Dorset and one sketch shows a number of the Foot family and nine others sitting down for a dinner party in the Chasm with all the men, except for a tree climbing youngster, in top hats. Others show dramatic peaks drawn earlier in the day with one including the path into the chasm and another the pinnacles at its eastern end.

Another year on, the German artist, Cornus, accompanying the King of Saxony, described *"the wonderful labyrinth of ruins"* but Walter White, on his way to Lands' End in 1855, provides more detail. *"The view of the great Chasm from below was not less impressive than from above and to wander about the confused masses at the extremity, to creep in and out of the caves, or climb to some of the little tables of turf on the top of the pyramids, was by no means an uninteresting experience, especially as you may learn a lesson in geology at the same time. It is here while looking into the perpendicular walls that you become aware of the tremendous nature of the subsidence, and you begin to fancy that perhaps it may be repeated before you get away"*.

In 1859, Murray's *Handbook for Travellers in Devon and Cornwall* was dismissive maintaining that the great chasm will probably disappoint as it too much resembles a gravel pit but by 1898 Mumford's *Illustrated Seaton, Beer and Neighbourhood* told of change, as the bare ruins of 1839 *"are now clad with vegetation of half a century and the mossy turf on the many broken terraces is studded with primroses in such luxurious bloom as to carpet the ground with a carpet of yellow"*.

Two years later the prolific writer Sabine Baring-Gould described a more aggressive coloniser in *A Book of the West*. *"The whole labyrinth of chasms is not to be ventured into with good clothing on, as the brambles grow in the wildest luxuriance and are clawed like the paws of a panther"*. In 1911 Francis Bickley, in his book *Where Dorset meets Devon* thought it *"must have been a ragged and raw place on the morrow of the cataclysm but today the Dowlands Landslip is a beautiful glade between land and sea, covered with grass and shrubs and trees. The trees show how fertile a country this is"*.

In 1939, Dowlands faced a new threat as English Holidays Ltd bought a large section of the Undercliff to make the biggest holiday camp in Europe. Not surprisingly, there were plenty of objections as reported in the *Western Morning News* of 2 and 26 August when people's thoughts might have been elsewhere. War had been declared, the West Indian cricketers had gone home with seven matches still to be played, but even so objections raged. The Stedcombe Estate feared that *"a holiday resort would seriously depreciate its value and interfere with its amenities"*. The headmaster of Allhallows feared that *"among visitors to the camp there would be persons with no respect for private property or the amenities of the school"* and he was also afraid of the increased rate of disease. Further afield, Mr Pitt in Lyme Regis feared that *"if the Landslip became the front garden for 5,000 people during the season, with cafes, chalets and slot machines, the amenities of Lyme would be affected"*. Others feared that it would affect the Lobster and Prawn fishing, that it would be a national and not merely local disaster, and that *"if a thing was bad, it did not cease to be bad because it was going to help the rates"*. Some objections were more startling like the claim that *"from a geological point of view it was quite obvious that sooner or later the proposed site would slide into the sea. The weight of the camp must necessarily accelerate any slip"*.

In 1939 Muriel Arber wrote a short article in *Country Life* hoping that *"these cliffs and undercliffs may be preserved from the exploitation with which they have been threatened"*, and in 1940 she wrote *Coastal Landslips of South East Devon* marking 100 years since the Landslip. Long after the war, Norman Barns explained how the Undercliffs survived the threat describing how *"they had problems with access, ingress and egress, and while they were trying to sort that out, the Nature Conservancy moved in"*. They bought 132 acres of Dowlands Undercliff for £1,000 in 1953 by which time the Conservancy had completed negotiations with landowners before the declaration of the National Nature Reserve in 1955.

By then, Muriel had resumed her annual April holidays in Lyme, always staying in the same room at the Alexandra Hotel. Many years later, Nicky and I came to know her through contacts with John Fowles and the Philpot Museum. When visiting us in Combpyne she recalled the first time she had paid her six pence in the kitchen at Dowlands before going down into the Chasm. We fantasised that Professor Buckland, so quickly on the scene of the Landslip, had actually stayed in the house where we were having lunch. It had been the home of Mary Oke, sister of Elizabeth, Buckland's mother. At that time, we did not know that Val Baker, who had previously lived in the Manor House, had sent John Fowles the information that not only was it the house where Elizabeth grew up, but it was also where William Buckland stayed at times.

Muriel recalled the excitement of her first visit to Goat Island *"a Red Letter day in my life"* with Tom Wallace and warden, Laurie Pritchard,

The Great Landslip

William Buckland stayed in the Manor House, Combpyne

Taking the train to Combpyne Station, 1905 and January 1965

Landslip Cottage, photographed by Muriel Arber in about 1928

57

on 21 April 1958. She had written the *Country Life* article because the anniversary seemed to call for some account and, changing subject, she recalled the good condition of Underhill Farm when a scene from *The French Lieutenant's Woman* was being filmed there. My book celebrating 50 years of the National Nature Reserve is dedicated to Muriel *"who loved the Undercliff and provided perceptive accounts of the geomorphology"*.

In 1988, she recalled her memories of early Undercliff visits in *Lyme Landscape with Figures*. *"As the footpath leads into Devon, the county boundary used to be marked by double gates; the path was often flooded at this point. Beyond Ware Lane, in old days, Rabbits kept the Undercliff reasonably clear. Donkey's Green was then an open space with a grassy mound near it known as the Giant's Grave. Ames' wall at Pinhay was still standing on either side of the path; halfway along it one could emerge onto the grassy slopes of Pinhay Warren overlooking the sea. Beyond the wall the main path led on for a stretch that was always carpeted with pine needles and deliciously springy underfoot. There was a little plateau, hidden by bushes, seaward of the path, from which one got a marvellously peaceful view back to the eastern side of Pinhay Bay which seemed remote from everything in the world except birdsong and the sound of the sea. One could easily climb down to the bay by the steps, which have now been buried in a mudflow. At Pinhay the path descended to a valley that echoed to the beat of the ram that pumped water to Pinhay House. Soon, one was within sound of the chimes of the clock on the castellated mansion of Rousdon"*.

Muriel and friends distinguished *"between the Undercliff, within easy reach of Lyme, and the Landslip which is, in fact, the Undercliff on the seaward side of Goat Island. We were not then aware of the existence of Goat Island and the Chasm. When I was still too young to walk the whole length of the path, we went to picnic each year at the Landslip, taking the train to Combpyne station and walking thence by lanes to Dowlands Farm. At the farm we bought our six penny tickets and followed the track down to the Undercliff. We ate our lunch there and then wandered about for the afternoon before having tea at Mrs Gapper's cottage. Rough wooden benches and tables stood under the trees and Mrs Gapper's daughter, Annie, came to take the orders and returned with trays laden with tea, bread and butter, jam, and cake, all of which she must have fetched from Lyme or Axmouth several miles away along the paths. I have never seen so many Chaffinches as there were in that cottage* *garden; they hopped on the tables and flew about everywhere undeterred by the presence of Mrs Gapper's white cat. After tea we made our way down the rough path to the shore, a place of sublime peace. Here, I scrambled among the great boulders of fallen Greensand and slabs of Lias".*

Muriel was to have new Undercliff experiences much later and, therefore, links the pre and post war enthusiasts. Among the latter, Norman Barns and Tom Wallace were prominent and John Pitts must have known the slopes, ridges and hardly penetrable bramble at least as well as them as he worked towards his PhD thesis. Another geologist, Keith Moore, knew almost every inch of the Reserve's shoreline and later, Ramues Gallois and Richard Edmonds, developed their sometimes contrasting theories about the Landslip. Ramues wrote accounts of each of the geological periods relevant to the Undercliff for Natural England, and is an expert on the World Heritage Coast. He has clambered on the coastal cliffs to get first-hand knowledge of individual rocks and strata eventually deciding that Buckland and Conybeare were *"almost right first time"*. Richard, aided by new technology, old texts and the same enthusiasm as Ramues, has developed ideas and profiles which are summarised a little later.

Between the 15th and 23rd January 1840 William Dawson made the measurements which are shown on a *Ground plan and Section of the Great Landslip*. Access to the point of the first measurement became possible again in 2021 after clearance of dense scrub at the western extremity of Goat Island. At that point the gap between the new "Island" and the old cliff from which it had moved was about 200 feet and the cliff height some 130 feet. From the base of the cleft, just off the old coast path there are two possible steeper routes up to the present coast path some 200 yards distant. It has come down from the farmland above by 50 steps before following an indirect route to the steps that lead up to Goat Island. The direct route from one set of steps to the other would be just over 100 yards. This must be where Dawson measured his second section as 350 feet across.

There is a way into the Chasm from the base of the first steps and Dawson's third measurement was some 200 yards along this route at a point where an old hedge bank comes down from Goat Island. The cliffs were only 300 feet apart with the inland one 105 feet high. At his fourth point, 250 yards further on, Chasm width was 363 feet and as the "path" and the base of the Chasm had gone down a slope, the near vertical inland cliff is, as it was at that time, some 210 feet high. Another 200 yards on and everything is more chaotic, the gap at 430 feet is wider

Chapter 3

and the inland cliff far less steep. On the other side there are several obscure routes up to Goat Island. Slightly further on to the east the land falls away steeply among spectacular rock pinnacles which have been much eroded since 1839. Dawson's fifth section continues down to the newly formed beach some 575 yards away where a ledge of rocks had been forced up from the bottom of the sea, varying in height from 20 to 50 feet.

One of Richard's reinterpretations, *Goat Island and the Chasm Explained*, pays particular attention to Conybeare's cross-section which was based on Dawson's measurements. Richard maintains that to understand a landslide, one needs to start with a map and from that to produce a model which is just what Conybeare and Dawson did. There was, however, a problem, for in January, Dawson only drew one complete section, from the clifftop to the beach. This just clipped the eastern end of Goat Island. Conybeare's key section shows the 'new raised beach' at the right-hand end. That beach is to the south east, not the south, of Goat Island, so the illustration is not the north - south section one might have thought it to be. Richard suggests that the diagram is a 'composite' made up from the different sources and emphasises *"that the majority of beach uplift (is) in front of Dowlands, not Goat Island"*.

1806, 1981 and a recent version of the Bindon and Dowlands landslips.

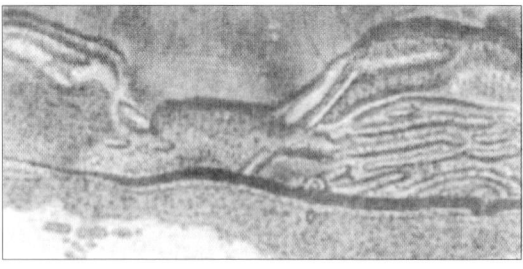

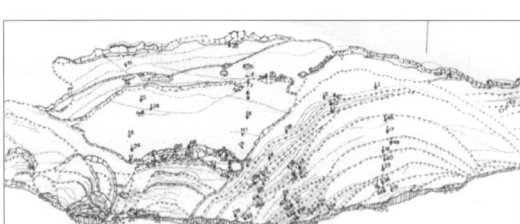

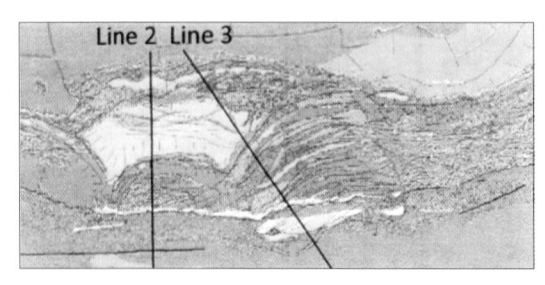

Of critical importance, but perhaps not recognised until the work of John Pitts, is that Bindon is not the only site of movement, for an earlier Dowlands slip is critical. Its arcuate ridges and depressions were shown by Searle in 1806 and it was Dowlands that failed on Christmas Eve and throughout the following day. The movements there *"took away the support for Searle's cape-like projection"* and when it started to move, the formation of the Chasm and Goat Island began. As the new

The Great Landslip

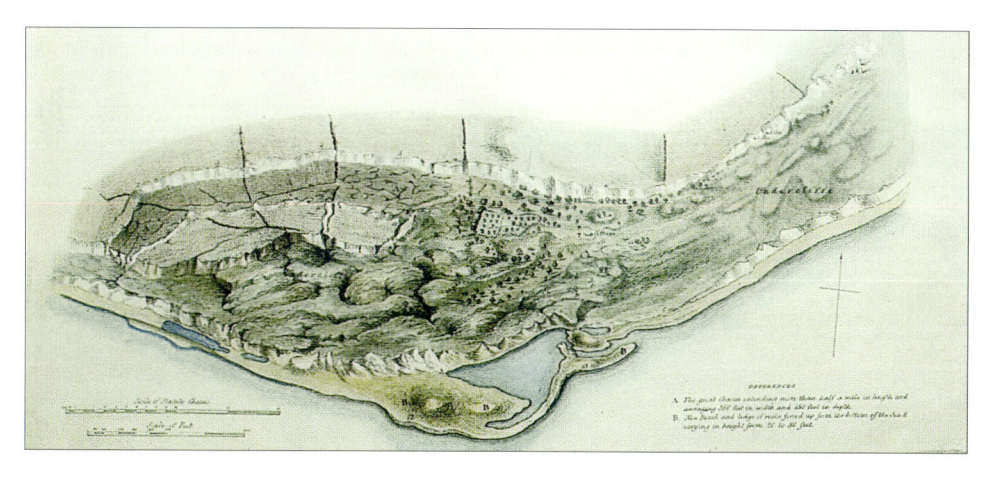

Ground plan of the Great Landslip from Dawson and Conybeare's 'Ten Plates', 1840

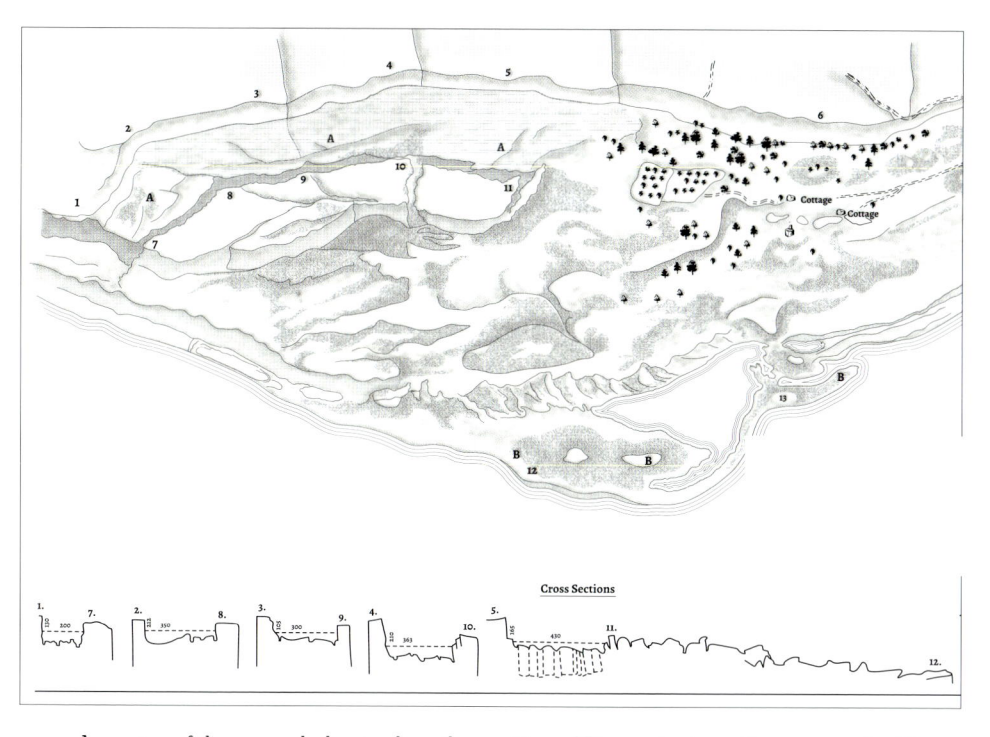

A version of the ground plan to show the position of Dawson's six sections across the Chasm together with his measured profiles.

Chapter 3

'island' slid seaward, the Chasm rift deepened and blocks from both sides collapsed into it. The series of cliff falls ended with the collapse of the main block of the inland cliff which came to form the floor of the western end of the Chasm. The back of Goat Island also continued to slide 'into the void' and because of the slight rotation, the failures at the eastern end of the Chasm were as we have seen more complex than any in the west.

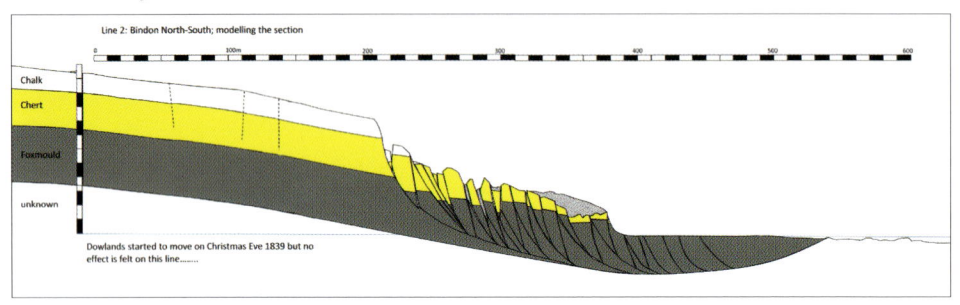

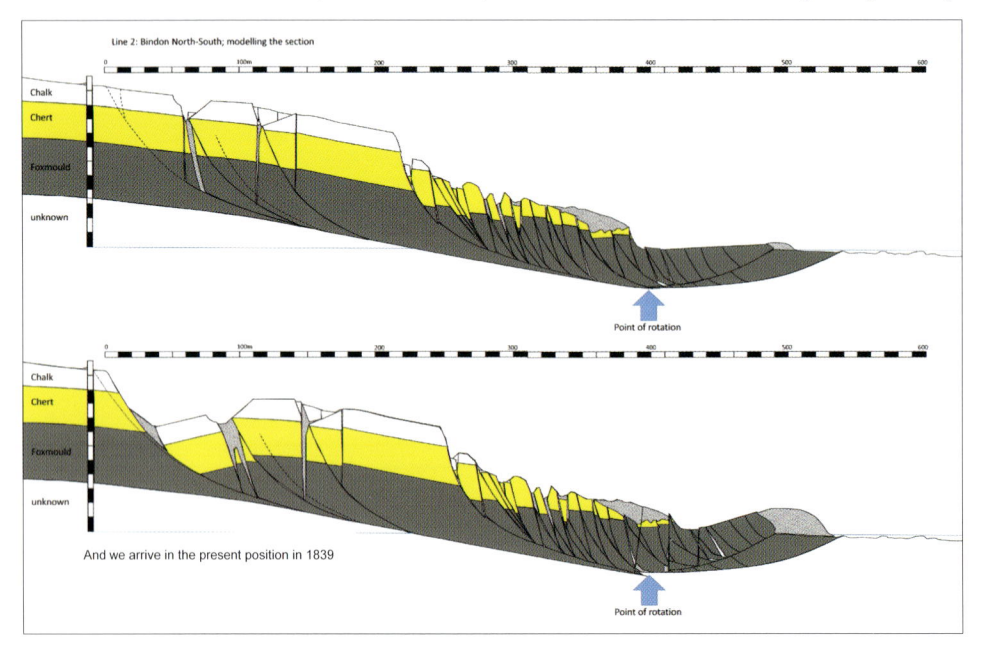

Richard had accumulated enough data to attempt a cross-section based on Plymouth University's Lidar data which could penetrate to the base of the chalk and then, using knowledge of the position of the chalk, theorise about the position of the unconformity. This indicated that the strata are folded, as elsewhere, and that the unconformity drops away below the line suggested by Conybeare. Richard then envisages three stages in the process of Chasm formation and Goat Island movement.

Dowlands failed first, on Christmas Eve 1839, taking away support for the headland that would become Goat Island. As that support was taken away from the south and east, Goat Island moved by rotating in a horizontal, clockwise sense, having the effect of unzipping the Chasm from the east to the west. So, early in the Christmas Day failure, we would have seen a deep fissure opening up in the north-east, behind Goat Island, but that gap would have narrowed to just a crack in the western end. As this widened it created two vertical cliff faces, one behind Goat Island, and another in the landward cliff face. A series of massive blocks failed off the back cliff face and the seaward faces of these also collapsed into a deep rift that was forming as Goat Island slid seaward, towards the south and east. The entire back of Goat Island also slid northwards into this gap, filling the deep rift with collapsing scree material.

A little later the massive Chasm block slipped down the landward cliffs to form the back tilted floor of the chasm. The ravine on the western side of the Chasm is much narrower because this is where Goat Island has simply slipped away in a south-eastern direction, allowing a few blocks to slide into the rift. The main beach heave is in front of Dowlands and to the south and east of Goat Island because this is the main direction of movement.

Goat Island may have moved about 35m to the south and east. Dowlands may have moved about 10m before Christmas Eve and Day before it unloaded, and took away the support, for the headland.

The underlying structure is probably a shallow fold of rocks very similar to what can be seen in White Cliff between Beer and Seaton. Goat Island lay on the southern side of this fold, primed to fail spectacularly when erosion inevitably took away the support. Movement possibly stopped when it did because the angle of the dipping rocks decreased towards the axis of the fold.

Another version of the event was recently found by Undercliff enthusiasts Roger and Kath Critchard, in the archive of the South West Heritage Trust (Ref 5439 Z/F/1). The account is in a letter from one of the sons of the Rev Frederick Barnes, formerly vicar of Colyton. It was probably the same son who gave the account of four of the family going down into the Chasm two years later. (See chapter 4)

Chapter 3

"My dear father,

"I began a letter to you ... intending to give you some account of the landslip at Bindon ... If Dr Buckland has returned to Oxford no doubt he has given you a much better description than I can.... We first went along the top from the centre to the east till we could descend and then proceeded towards the site of the cottages where we soon found symptoms of the great convulsion in the cracks across the path. The first cottage is excessively shaken, the walls still stand but the windows are thrown out and the floor of the cottage, sand and lime concrete, has been lifted up almost bodily nearly halfway up the fireplace ... Our next stop was to take a view of the chasm or hollow immediately under the ground that still keeps the line of the top. This may be about 70 yards across in parts much broken up with points of the rock standing up as at Pinney and this, I should think extends east and west nearly three quarters of a mile in all. It must have gone down quietly I think at first as the surface is now not very much broken ... and the fissures and cracks have increased since it first took place, indeed are still going on a little, across this chasm stands what was the former cliff, the surface being part of the same turnip and wheat fields of which part are now in the hollow, part in their original position.

"I met young M... of Colyton he was there the day after, rabbit shooting but remembers observing some cracks in the path, he then clambered up the high separated mass and says that when he just got up the surface of the turnip field was nearly in its natural position but while there it opened in some places and rose in others. Some of his party lying down through fear, they then hurried back and found the windows of the cottages out and the fissures in the path wider.

"We had not time to trace our way all up the chasm, but went around by the face of the slip towards the sea. Unfortunately the tide was coming so that we did not see the whole extent to which the coast had been protruded seaways. I made the attempt and succeeded at the expense of a wave clean over me in going round the artificial Bay. The stones nearest the rock are the original beach lifted up, while the bank to the eastward that forms the breakwater of the artificial bay is composed of the stones and rubbish forced up from under. These will doubtless soon be washed away, as the mould that holds them together is chiefly the greensand. Much of this bank was washed away while I was even looking. I was well repaid for my trouble and ducking as it was the most convincing proof to me of

what had taken place. I then, for nobody ventured to follow me though they might have escaped between the waves perhaps, went along the face of the rock and took a view of the chasm eastward. I wish we had had time to ascend the still erect part, but we were obliged to go on and ascended the rocks towards Mr Hallett's, where we proceeded home ..."

He did not call the ground on which the cottages stood to be part of it but thought they were only shaken down. He didn't know what caused the first move but the fact that the ground has been raised a long way out to sea and that the bank is composed of the same materials as those beneath the surface rock convinced him that the soil that had filled the vacuum had been forced out. That was confirmed by the fact that far out to sea, where the anchorage used to be stones, it was now mud. The writer feared that his father would make neither head nor tail of his account. *"It would be a difficult matter for anybody to make a sketch of it"* but that is just what he attempted.

An interpretation of Barnes' original sketch, seen to the right.

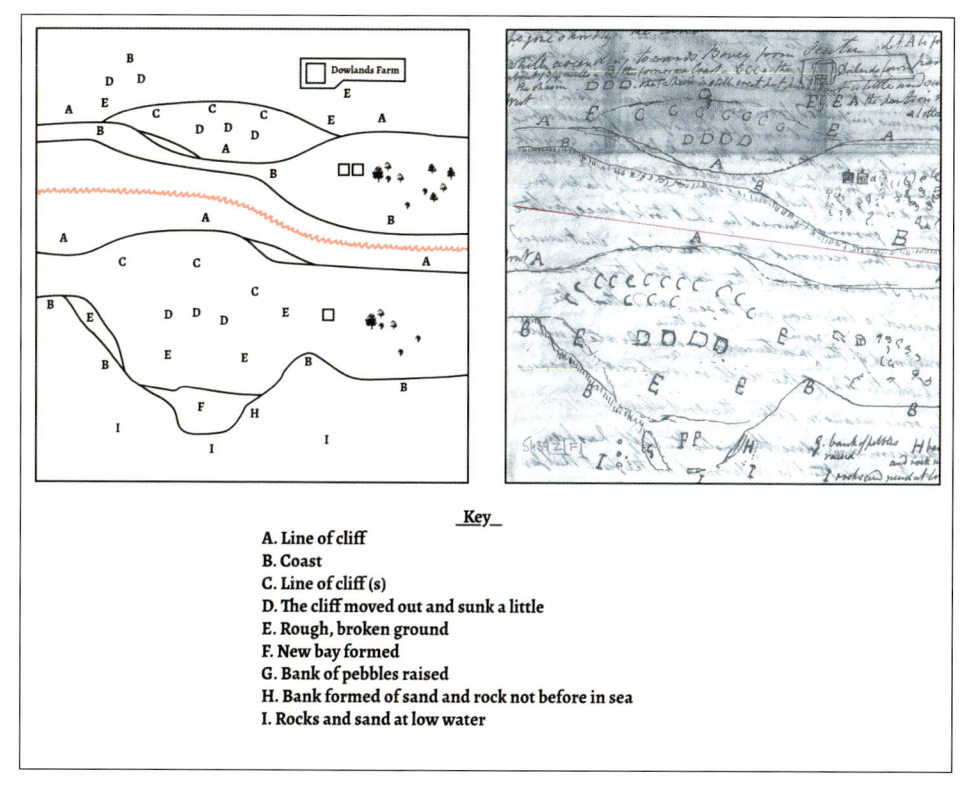

Key

A. Line of cliff
B. Coast
C. Line of cliff (s)
D. The cliff moved out and sunk a little
E. Rough, broken ground
F. New bay formed
G. Bank of pebbles raised
H. Bank formed of sand and rock not before in sea
I. Rocks and sand at low water

Chapter 3

Since counting birds around Goat Island in 1995 I had been taking groups through the Chasm meeting various difficulties. In 2020 the Jurassic Coast Community Coordinator, Guy Kerr, arranged a series of coastal walks led by volunteer Ambassadors. Guy joined the final walk of the series when I led a party through the Chasm to the Plateau and back over Goat Island.

Guy, who had travelled in the jungles of Costa Rica and Ecuador describes the Undercliff as *"Amazonian in nature"*. He had not expected the *"incredible otherworldly environment"* or such *"a jawdropping traverse"*. He wrote that *"as impressive as the reserve itself was the seemingly effortless aplomb with which the seemingly ageless leader hopped branches and scampered through crevasses accompanied by his faithful dog Fuggles."* I was later to appreciate some of the walkers' comments to Guy: *"really enjoyed the Undercliffs walk, unlike anything I had done before", "the leader was knowledgeable, passionate and enthusiastic ... it was a truly magical, exciting and unique walk"*.

I was reminded of these descriptions late in 2021 when Woody and I, having cleared the path of brambles and nettles for the third time in the year, continued slightly too far to the east reaching a point where the land suddenly fell away down an impressive precipice. We had only gone 50m too far to reach a spot that, after twenty-five years of exploration, was completely unfamiliar. Such is the nature of the place.

The Great Landslip

Goat Island and the Chasm

The lagoon that formed briefly after the landslip.

Both illustrations from rediscovered and cleaned images drawn in the summer of 1840

Monitoring Landslides

Investigating landslides is technically difficult. Analysis of aerial photography is often used but inadequate where there is abundant vegetation. Laser reflectors and satellite GPS systems have also been used. In the Undercliff a terrestrial photographic technique has been used by Exeter University to compare the same location over a period of time. For example, at the pumping station in February 1994 photographic evidence showed that after heavy rain there had been movement of 3.4m at the toe of the landslide in a week.

April 2003

April 2004

May 2005

A GPS survey station used to measure movement.

Movement Diagram showing forward displacement by Grainger (1995).

Fixed point photography. South West Water engaged Peter Grainger from Exeter University to investigate the problem of the future of the pumping station.

CHAPTER FOUR
Landslips

By virtue of its position, Allhallows was always likely to attract teachers with an interest in the natural world. It had certainly been a factor in Tom Wallace's career change and presumably so for geologist Keith Moore who had been teaching there for 25 years before becoming Headmaster. Two years before I interviewed him in 2001 the College had suddenly been closed when finance was withdrawn but he had secured an alternative position at Taunton School. My first question was about his use of the Undercliff when teaching.

"I used it as much as I possibly could, not only with Allhallows students, but also with adult education groups, and Bristol University extra-mural classes. Every opportunity to go into the Undercliff I took. One of the things that was most noticeable was that it was never the same two days running, minor movements, a bit of cliff fall, the path not quite the same. I wish I had documented much more of what was going on".

When South West Water were concerned about the status of the Pinhay pumping station, they engaged Peter Grainger from Exeter University to record events using fixed point photography. His problem was finding a fixed point. Eventually he selected a large 5 metre Greensand boulder on the beach as it was about the only thing that wasn't moving. Keith remembered surveying the slip in the mid-seventies when it swept away the last remnants of the route from Rousdon pumping station to Humble Point. It was the path the pebble pickers had used in the sixties.

I next suggested that he could talk me through the Undercliff section of the pages illustrating the geology of the potential World Heritage site. These were in the Nomination Book putting forward the case for this prestigious status. Keith said *"It is a superb document; the only sad thing is that its circulation has been so limited. It shows what*

Chapter 4

you might see with all the greenery removed. It is immediately obvious that there are two parts, the lower Triassic and Jurassic material, separated by the plane of unconformity from the mainly Cretaceous material above. That is the key to understanding. The lower rocks are quite extensively folded and faulted, not very dramatically but enough to make them harder to interpret. In the 25 years that I was involved in the Undercliff, nearly all the lower groups of rocks were exposed somewhere, at some time, except where the slip at Humble Point obscured the geology and where the subsidence of the Plateau brought chalk down to sea level."

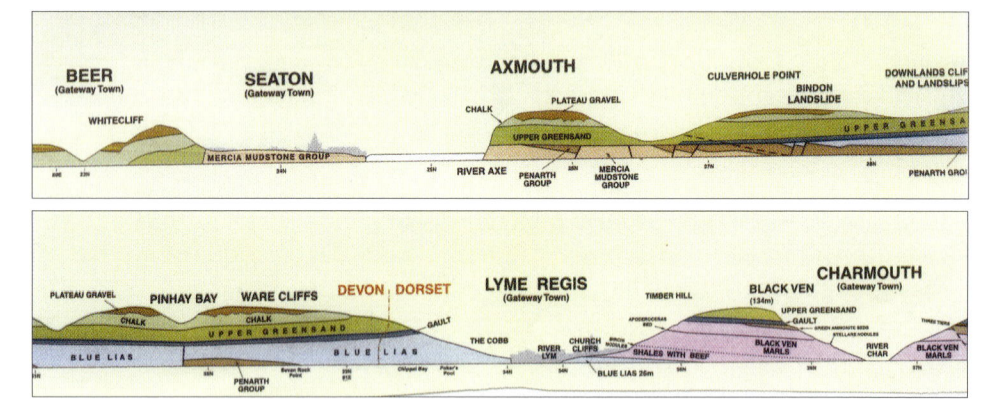

Sections of the Undercliffs stretch of the World Heritage Coast
From the Nomination Handbook

The division into two structures is very important as the lower rocks, with their folding and faulting, are lower Jurassic or pre-upper Cretaceous. Mid-Mesozoic, mid-Jurassic movements, precursors to the Alpine event, produced the faults which contrast with the higher upper Greensand and Chalk which lie undisturbed, virtually flat with little folding. Two of the faults which do occur, penetrate the Greensand and Gault and, indeed, the Chalk, and it is interesting that these are not shown on the section. After a few words about maps, the representation (or not) of faults and the value of constant monitoring. I then commented on the need for regular observation.

"That is right, it is a matter of luck and chance. As I say, I have seen the entire exposure of the Rhaetian succession at Culverhole, the complete section, but never all at once. Observing over the years I have been able to piece it together just by the chance occurrence of being in the right place and time. The lower section is the most

revealing where there have been exposures because of landslides and marine erosion".

Since that time, marine erosion has been extensive, and landslides were particularly prevalent in the subsequent winter. Because of the foot and mouth outbreak and the closure of the coast path, I had been on the beach more often and been surprised that with the pumping station gone and water being discharged into the lower undercliff, the toe of the slip had hardly moved and the coast path there had not been disturbed. I suggested that there were a variety of different causes of slips, that what happens at Haven Cliff may well be different from what happens at Pinhay.

"Absolutely right. I think there is a tendency to assume that along the whole of this coast from Branscombe to West Bay that any failure of the cliff actually is the same mechanism in all places, and it certainly is not. Black Ven slip is a very different structure from the sort of thing we have got at Humble Point or Goat Island or, indeed, Haven Cliff; it really depends upon which of the underlying strata are involved. The Blue Lias and Mercia mudstone group are relatively strong compared with the higher groups of Jurassic that underlie Charmouth and further east, so there is a tendency for mud slides to the east, in the lower group of rocks below the unconformity. This is almost unknown in the Undercliff itself".

When I ask for an explanation of the obviously crucial unconformity Keith explained *"it means a break in the conformable sequence of strata. The reference to it here is that Blue Lias and Mercia mudstones beneath that plane are disturbed, folded and faulted. A time break in sedimentation then occurred after the lower group of rocks had been deposited on the sea floor (or in the case of the Mercia mudstone group in lagoons and coastal lakes above sea level) and for a period of time subjected to folding and faulting and then to erosion before the whole area was resubmerged for the deposition of the Greensand and Chalk and then further uplift, without much in the way of distortion, to show the cliff as it is at the present time. So, the unconformity represents a time break and if one took the middle of the section, say, at Rousdon or Charton, the Blue Lias was laid down some 180 million years ago and the Greensand 100 million years ago. Thus, the time break is of the order of 80 million years. Unconformities were first recognised in Scotland in the nineteenth century and were one of the features which were fundamental in the almost heretic suggestion that things have not always been as they are now".*

Chapter 4

More questions followed with a final one about changing names; the Tea Green Marls having become the Blue Anchor formation and the Rhaetic becoming the Penarth Group. *"Standardisation is the reason"* explained Keith. *"As the geological map evolved, workers in specialist areas tended to give rock groups local names, just as the Lyme quarrymen did. Over the years, therefore, there has been an attempt to standardise the lithostratigraphic rock groups to match the fossil based Paleostratigraphic group. They do not necessarily match up as organisms evolve at varying rates which are independent of the rocks, which depend on local deposition conditions. Ammonites lived in surface waters and when they died and were transported downwards, they could end up in any marine deposit."* Finally, Keith told me that a stratotype *"is a type of stratum, recognised at a particular site by its fossils, and taken as a type example for comparison with other areas"*.

Peter Grainger was mentioned in the context of landslips, the pumping station and of articles he had written in the Proceedings of the Ussher Society. In 1985, Peter went back 50 years to the time when a spring of clean water, Hart's Tongue Spring, which issued from the base of a large, slipped block of Chalk and Chert about 30 metres above the high tide mark and 160 metres from the beach, was tapped by the South West Water Authority, becoming the sole source of water for Lyme Regis. A pumping station in front of the spring sent water inland to a small reservoir in Rousdon.

He then explained *"the basic cause of the large infrequent landslipping events and the continual smaller movements, is the presence of weak rocks in the middle of the geological succession coincident with a zone of high ground water pressures and seepage towards the coast. The Cretaceous aquifer of Chalk and Upper Greensand overlies Triassic and Jurassic strata dominated by impermeable mudrocks. Under-cutting of the base of the cliff then removes the debris deposited there by landslide activity"*.

Apart from Grainger's persistent observations and recordings from 1980 until the final abandonment of the pumping station, it is convenient that the Allhusen family have recorded rainfall at Pinhay on a monthly basis since 1868. As water takes time to move from clifftop to the 'impermeable mudrocks' below, rainfall in the time before a slip may be relevant and is frequently included in the data in the coming pages. Tide and storm damage events at the cliff base can be equally significant. Only landslips in the 150 years since this recording began are considered. The average rainfall over that time has

Monthly Rainfall

Chart 1 – Contrasting rainfall in different months 1870-2019

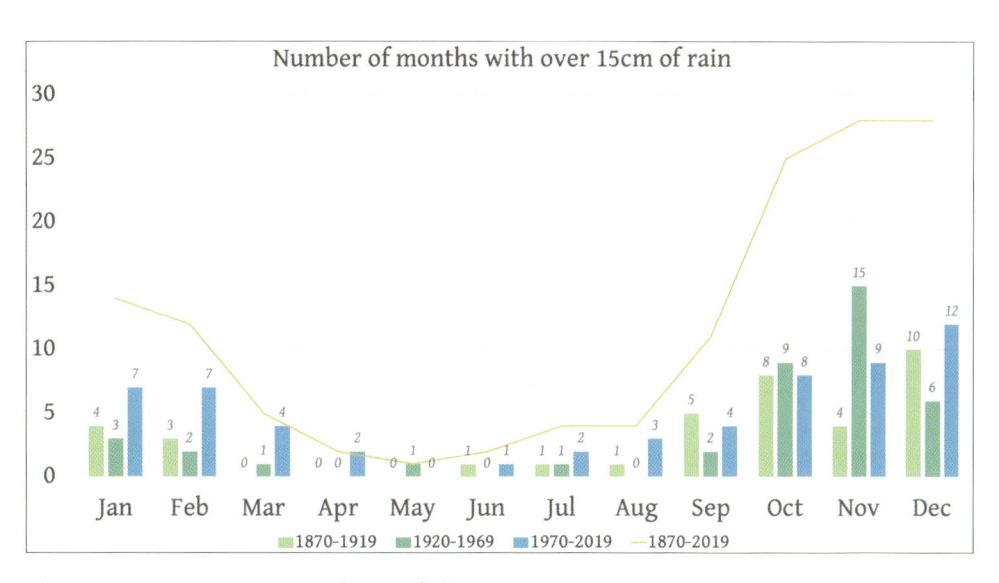

Number of months with over 15cm of rain

Month	1870-1919	1920-1969	1970-2019
Jan	4	3	7
Feb	3	2	7
Mar	0	1	4
Apr	0	0	2
May	0	0	1
Jun	1	0	1
Jul	1	1	2
Aug	1	0	3
Sep	5	2	4
Oct	8	9	8
Nov	4	15	9
Dec	10	6	12

Chart 2 - Average Annual Rainfall 1870 - 2019

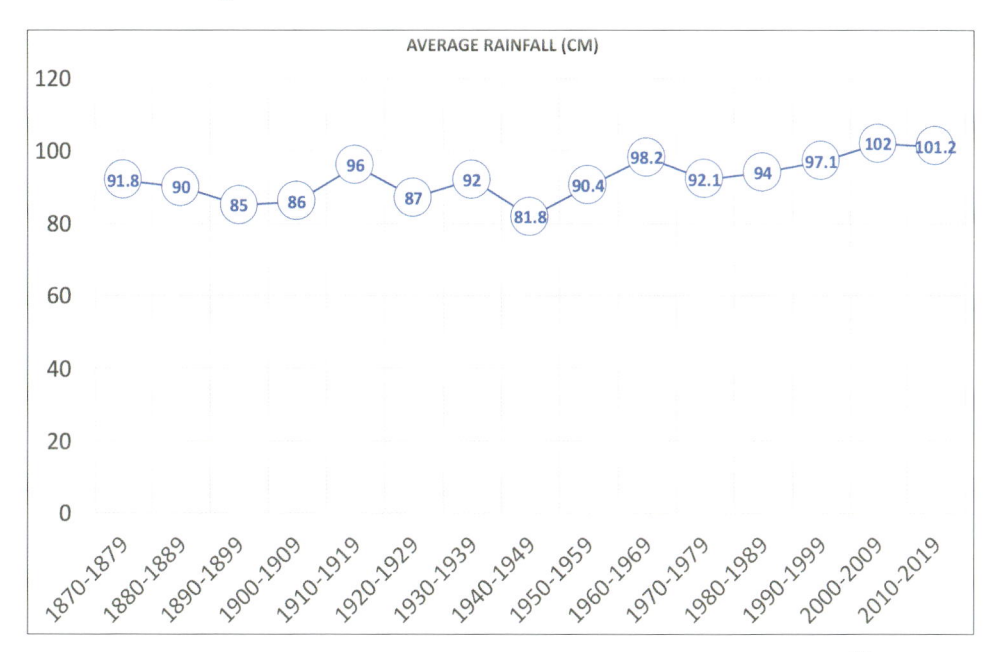

AVERAGE RAINFALL (CM)

Decade	Rainfall
1870-1879	91.8
1880-1889	90
1890-1899	85
1900-1909	86
1910-1919	96
1920-1929	87
1930-1939	92
1940-1949	81.8
1950-1959	90.4
1960-1969	98.2
1970-1979	92.1
1980-1989	94
1990-1999	97.1
2000-2009	102
2010-2019	101.2

Chapter 4

been 7.8cm a month. The information about falls come from a variety of sources, as indicated, but some may have been missed.

Peter Grainger had been monitoring events for English Nature since 1980 and when I interviewed him at his home in Exeter in August 2001 the wet months that were to cause so many slips along the Undercliff were about to begin, but even the previous winter had led to movements at Pinhay, larger than any since 1985. When I asked about the different mechanisms involved, he answered *"It is rather like earthquakes; you can have lots of small events happening frequently, but statistically it evens out as in one area quite a lot may happen, but in another area, things get locked up, as it were, and then everything gets released in a very large movement very occasionally. What we saw last winter was due to mechanisms involving rainfall frequency and the heaviness of individual rainfall events rather than any effects of the sea".*

"What tends to happen is that particularly heavy individual rainfall events trigger landslipping where that is sensitive to extra water coming in at the surface. The deeper, larger landslides tend not to respond to individual events but to months, or even one or two years, of high groundwater levels. Last winter it built up enough by January, when events really got going, but it was rain that had come in October or September building on groundwater levels that had not gone down as much as in previous summers".

I ask whether he has any fears or predictions about the toe now that more water is being released there. *"Yes, indeed. When the pumping station was not extracting water, which it did not do in the winter, the surplus was piped to the beach and the pipes were reasonably maintained. Also coming down to the beach is the Whitlands stream which is piped to prevent problems on the path higher up. In front of the pumping station the main pipe is discharging as a waterfall which will keep water levels up so the area is continuously active. My report to English Nature in July suggests that last winter there was 6-8m of forward movement. The sea keeps pace with that, so it looks similar from the foreshore. The pattern of the coastline is that the beach position is not eroding inwards; it is like a glacier".*

Inevitably, I want to hear more about Peter's methods. *"It is mostly observation and traditional surveying. In the early 1980s, the South West Water Authority, as they were then, was already measuring across wooden stakes over cracks. They had a system of physical surveys by tape and levelling. They also put in*   [To p. 79]

Landslips

Sam Edwards' Cottage as it was before the collapse. (See p.76)

The Great Cleft above West Cliff Cottage, about 1895. (See p.76)

Chapter 4

Table 1 – Major landslips 1876-2014

1876	A fall at Pinhay which destroyed East Cliff Cottage "after a thaw" but no dates mentioned (quoted by Norman Barns). There had been 37.3cm of rain in the previous October and November.
1879	Ten acres near the mouth of the Axe "slip into the sea" (*Sheffield Telegraph* 10/11/79). 60cm of rain between June and October inclusive.
1886	The "great cleft" at Whitlands widening. 18.3cm in December.
1900	The "great cleft" still widening (Woodward and Young). 17.6cm in January.
1911 October 9	Sam Edward's cottage, below Rousdon, falls some 10m after only 2cm of rain in three months from July and 11.2cm in the first 6 months of the year; so, it could be the drying out of the Greensand that caused the collapse.
	The next fifty years had no significant recorded falls and were relatively dry averaging 7.4cm a month.
1960	A fall of some 5m along the whole clifftop continued to develop until March 1961 (Barns) after 139cm of rain in a wet year.
1960 November	Movement below the pumping station and step slipping at Culverhole (Wallace). 12cm a month from August to November.
1961 January	Step slipping at the east end of Haven Cliff (Wallace) with crevasses nearly 4m deep (Arber 1973). 14.3cm a month from November 1960 to January 1961.
1961 February 28 – March 7	The cliff east of Humble Point falls 7m after 23cm of rain at the start of the year (Pitts) which leads to upheaval of 3-4m of Lias clay on the foreshore (Arber).
1961	Deep crevasses and miniature chasms on clifftop at Bindon, map reference 268 897 (Arber).
1961 – 1962	The steps down to Pinhay Bay collapse (MacFadyen). 25.2cm of rain in December and January.

1963 – 1964	Significant falls at Lyme and Pinhay (Barns) perhaps linked to the 23.4cm of rain in November 1963.
1968	Footpath "broken" 1km into the Reserve from Ware so that the path was moved further inland (Arber).
1968– 1969	Ware Lane "broken" just east of entrance to the Reserve and the only house there, The Orchard, had to be demolished (Arber). The house was replaced and has recently been replaced again.
1969 June 22	The track down to Humble Point destroyed after 14m displacement (Arber and Pitts). No more carriages or cars to the beach.
1976 – 1977	Extensive movements at Pinhay Warren, Rousdon and along Haven Cliff (Pitts). 16.3cm a month between September 1976 and February 1977.
1979	Subsidence east of Humble Point and from Charton Bay to Axmouth (Barns). 114.9cm in the year with 18.4cm in January.
1981 August	Extensive falls below Pinhay Warren

Extensive falls below Pinhay Warren. August 1981.

Chapter 4

1984	
February 22-23	Power lines disrupted as the pumping station itself moves slightly (Grainger).
1985, Early Spring	A large area of Pinhay Warren slumped some 25m despite, or because of, dry weather, with only 6cm of rain a month between March and May (Barns).
1986, Late winter and ongoing	Several hundred metres of path above the 1969 slip dropped and the inland cliff west of Rousdon continued to lose height (Barns). 32.9cm in December 1985 and the following January.
1990	
January 25	Several thousand tons fall at Haven Cliff moving material that had been unstable for some 80 years (Barns). 30.8cm in December and January.
1995	
February 25	Power-lines disrupted as the pumping station moves again between 20 February and early March (Grainger). 19.9cm in the first 2 months.
2001	
February	600m of coast path above Haven Cliff falls 2-5m but remains intact. Below some 30,000 tons of cliff falls to the beach and is soon removed by wave and tide action (Campbell). More than 85cm rainfall between August 2000 and January 2001.
2002	Movements of various types in most areas (Campbell). 17.8cm a month between October and December 2002.

Collapse of the coast path in 2001.

auger holes, for piezometers to measure water pressure in the ground. So, the methods then were to map the shape, the geomorphology, to measure across individual scarps and cracks with survey pegs. When we came back in the 1990s, we had developed photographic techniques at other sites so we could take a colour transparency from a previous visit, put it in a viewer on site, and view that transparency with one eye using the other to look at the new scene. We would then estimate, from that comparison, the changes that had taken place. If there was a small difference then you were able to move your position on the slope, you could move back to match the transparency quite closely and see how far you would have to move. This gave a rough idea of total movement, certainly when working in the toe. There is actually another paper we wrote about the technique. We also used the same method in 1994 along the road and track to cut down the time used on physical surveys, using photographs as yes/no techniques, to see if anything had changed."

Keith Moore had mentioned the difficulties with fixed point photography. The Exeter team *"had done some measurements at Pinhay and other toe areas when you can fix yourself on a shore platform at low tide, it does not move, and take views of the cliff and return to a marker at the fixed point another time. On the footpath or among the landslides fixed points are relative. In terms of technique, then, we use a combination of photographic monitoring, traditional surveying, and some innovative new methods."*

It was some time after the closure of the station that most of the fabric of the building was removed but all sorts of pipes and tubes of metal and plastic remained between the coast path and the sea. Tom Sunderland who had replaced Albert Knott as Site Manager of the Undercliff organised a little party of workers which, as far as I can remember, included Dave Palmer from the Countryside Service, Peter Youngman from the AONB team and David Allen and myself as Undercliff people. My main memory is getting very wet and very muddy and leaving a horrible pile of assorted pipeline remnants for Dave to take away when he could bring a vehicle down from Pinhay to collect the mess.

A student project in 2007 started at the base of the inland cliff above the site of the pumping station and took a somewhat tortuous course down to the sea. Elizabeth Hankey was hoping to relate the plant life to changes in the stability, soil and particularly slope, which were all inevitably interdependent. The tortuous route she took was forced

Chapter 4

1. A narrow ridge 120m along the transect.

2. The ridge obstructed by climbing Clematis.

3. Further obstruction at 190m.

4. Subsidence across open ground with the coast path behind (235m).

5. Woodland plateau south of the coast path (280m).

6. Holm Oak roots temporarily support the cliff edge (330m).

Stages along Liz Hankey's awkward transect.

upon her by the nature of the terrain which included narrow ridges, dense vegetation and fallen trees. On the steep slope below the inland cliff at the start of the transect the moss *Brachythecium rutabulum*, favouring rock and fallen wood, was abundant as, inevitably, was Harts-tongue Fern. After 25m the slope levelled off and Ivy started a long period of dominance with plenty of Dog's Mercury in the shade. 100 metres down, Liz was forced offline and onto a ridge to avoid hazardous slopes but, within days, the ridge was obstructed by a fallen tree. Dog's Mercury and Harts-tongue continued dominant, but tangles of Clematis often determined the route. Above the remains of the pumping station young Sycamore became the main obstruction. After 250m the transect crossed the coast path and then rose to a flat area with Pendulous Sedge and Wild Madder under Ash and Beech. Remarkably, this area was clear of Bramble. Before the transect reached the sea, it crossed disintegrating cliff edge, fallen trees and collapsed boulders. Sometimes the roots of a more mature tree, like the Holm Oak illustrated, would bind the surface for a time but collapse was inevitable. I am not sure that Liz achieved all her objectives, but she provided me with a flavour of Undercliff slopes from top to toe; a matter of some 360m at this point.

Transects at other points have been mapped for a variety of reasons and at different times. Some are shown at the end of the chapter.

Until 1976, a path at the western end of Haven Cliff had been well used but falls made the cliff all but inaccessible. Thirty years later a route of a sort was established climbing up to the line of the old path by means of crude slippery steps. In 2020, a flight of 170 sturdy steps were built and drainage channels incorporated so that workers for Natural England could more easily gain access to the 'shelf' and improve the path, by choosing a route that avoided too many ups and downs. They removed Buddleia and reduced scrub to provide more open grassland. The warm slopes suit Adders and the chalk pinnacles are fine for Peregrine Falcons and Ravens, while Fulmars often breed even though their cliff site is not directly above the sea.

In 2014, a spectacular fall some half a mile east of the steps that come up from the harbour, started with the collapse of waterlogged material after high tides and waves had undermined the base of the cliff. This destabilised everything above with falls of large Greensand blocks and loose debris. These buried the path, parts of which had already been lost in the initial collapse. The slopes above were covered in unstable rocks of all sizes while the surviving Greensand pinnacles appeared highly likely to fall at any time. By mid-summer the whole slope was

Chapter 4

covered in vivid blue as Viper's Bugloss flourished in the absence of competition and was a bonus for passengers on a Stuart Line cruise heading for Lyme Regis.

A rough path was established over the scree and then east under Finger and Thumb, beyond which the scrub is dense with a mix of Gorse, Blackthorn, Bramble and more Elder than elsewhere. Further on, an area with mature Ash and Hazel and large Field Maple and Sycamore trees indicates long stability and different geology, while the remains of a substantial wall suggest that this was once an important way along the coast. Natural England now keep the way open as access is vital for management. Without their work the scrub, together with Bracken and Nettles, would soon obliterate the path. Eventually, one comes to signs of the 2001 fall which had taken some 30,000 tons of cliff down to the sea while leaving the coast path intact but some 3-5m lower than before. Until 2016, the Haven Cliff route joined the coast path just beyond that landslip. Now, following a series of falls at unstable Culverhole, the coast path follows the cliff top until steps take it down into the Undercliff and over Goat Island. The old route remains, as access to Culverhole is needed, but on either side, undergrowth is rampant, and the few large trees leave plenty of light for the scrub and the Blackcaps. At much the same time the Crow's Nest, built under an unstable cliff above the coast path, at the eastern end of the reserve, was destroyed by another fall.

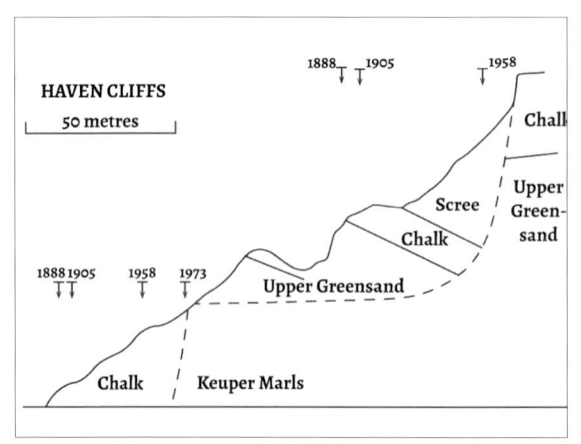

Facing page:

1. 1830 profile of collapsing cliff at Pinhay (H T De le Beche.)

2. H B Woodward drew this version of the results of the landslip.

3. A J Jukes Brown's profile of the same section in 1900.

Left:
John Pitts' 1981 diagram showing major changes to Haven Cliff.

Landslips

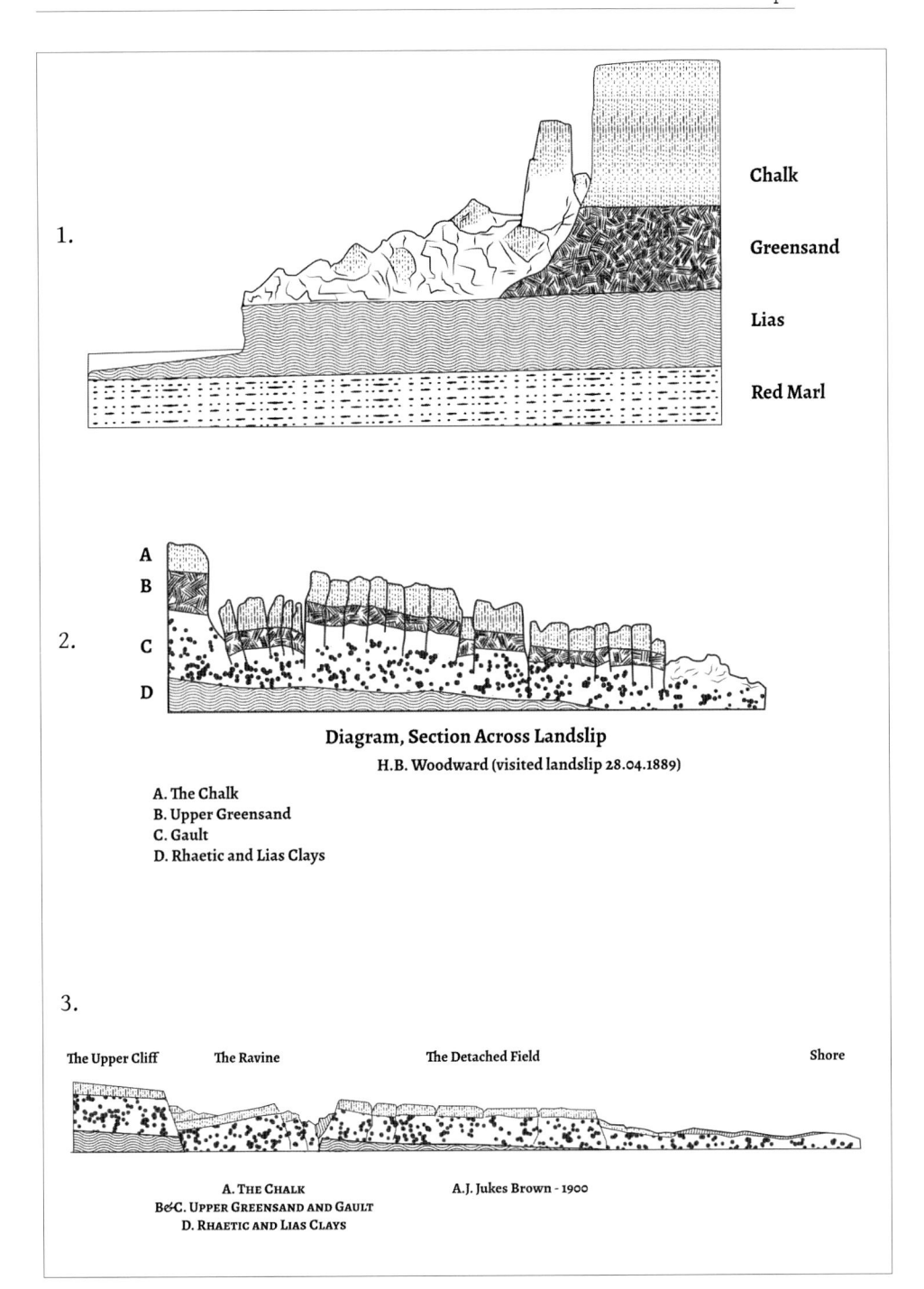

1.

Chalk
Greensand
Lias
Red Marl

2.

A
B
C
D

Diagram, Section Across Landslip
H.B. Woodward (visited landslip 28.04.1889)

A. The Chalk
B. Upper Greensand
C. Gault
D. Rhaetic and Lias Clays

3.

The Upper Cliff   The Ravine   The Detached Field   Shore

A. The Chalk
B&C. Upper Greensand and Gault
D. Rhaetic and Lias Clays

A.J. Jukes Brown - 1900

83

CHAPTER FIVE
The First 40 years as a National Nature Reserve

J F (Hamish) Archibald had a huge area to cover in his role as Assistant Regional Officer for the South West and must have had difficulties in producing the 1965 Management Plan. He admitted that without Tom Wallace's involvement *"the biology of the Reserve would be virtually unrecorded, and many sections of this plan would be extremely thin"*. Other authorities contributed ten pages on the local geology, seven pages on the soils and associated plants and brief comments on climate. The plant and animal life described in five pages was based on Wallace's work, as were the more detailed bird records, which formed appendix four. Hamish wrote the crucial section about objectives and the descriptions of maintenance work on fences and signs, as well as the fire precautions and details of public access. Four pages of bibliography referred mainly to the geological background but also included works by Nicholson and Tansley, as well as 15 references to Wallace and his lists of Undercliff wildlife. A crucial part, summarised in appendix one, described ownership, types of tenure, terms of agreement and legal arrangements which, renewed and updated, remain valid today.

The justification for his minimum interference policy is set out in the introduction *"Some of the Reserve is owned by the Conservancy but the greater part is founded on leases and agreements, which limit the Conservancy's freedom to initiate extensive management. In the event, management is not necessary as little could be done to increase the geological interest while the woodlands are in a thriving and regenerating condition – they include great numbers of exotic species, but these could not be effectively eliminated at this stage without a degree of disturbance which could not possibly be justified. In fact, the interspecific competition taking place and involving both native and introduced trees and shrubs is a most interesting aspect of the Reserve and, in certain parts at least, any management would be definitely undesirable"*.

Evidence of the disadvantages of the limited interference policy was clear in the section on soil and associated plants. The alkaline grassland, previously grazed by rabbits, was now dominated by aggressive grass species, particularly Tor Grass, and by woody shrubs which were also taking over areas of low scrub. Here, Bramble and Privet, often formed a continuous entanglement to a height of two or three feet. Some management was suggested for the scrub free parts of Goat Island; it could be divided into four parts, one of which would be cleared every three years giving a 12 year rotation.

There were also problems at both ends of the Reserve with dense Bracken in the east presenting a fire hazard while, in the west, picnicking and scrambling by trespassers in the cliff above the mouth of the Axe did not fit with the terms of the agreement with the Stedcombe Manor Estate. On the research front, Hamish wrote that visitors were impressed with the Reserve which led *"to a steady stream of research proposals but almost all are shelved as their initiators became conscious of the problems presented by the complexity of habitat and difficulties of access to, and movement within, the site"*.

Most Management Plans began by comparing the site under consideration with similar sites elsewhere, but the Undercliffs posed a problem as it was in many ways unique providing little scope for comparison. *"The original case for its establishment was based on geological and physiographical grounds. It was one of the most extensive and spectacular landslips in Britain and the only one in which major slips have taken place since the scientific study of geology began. As a result, geologists regard it as a geological reserve but there is little doubt that an overwhelming case could be made out on biological grounds alone. No strictly comparable woodlands are known and, though young, the Ash wood in the Chasm has developed in the almost complete absence of human interference"*.

Hamish's progress reports subsequent to the Management Plan were mainly concerned with local difficulties. In 1970, fire had burned 2½ acres of scrub below the track to Charton Bay and the track itself had been disrupted by landslipping. Trespass continued at the western end and the Bindon Estate complained about the state of the fence along the top of the Reserve. Next year, that problem was solved with the

Conservancy providing £500 for its replacement but not being responsible for its future maintenance. Other problems remained with the Bindon Cliffs lease. Mr Wallace's activity was restricted by illness but a footpath warden for the East Devon coast had been appointed by the Council. The 1972 report described damage to recently installed signs and failure to consult the Conservancy before permission was given for an operator to collect 750 tons of selected pebbles from Charton Bay. John Pitts was given permission to study landslipping for his PhD.

Nor were the Conservancy consulted about plans for a cable car to carry 100,000 passengers a year up Haven Cliff. Those plans, abandoned because of cost, were part of the 1971 Axmouth Study which hoped for extensive development in Seaton Marshes with a lake, 150 boats, 200 small homes and associated recreational and car parking facilities. A boat route through the shingle below Axmouth Bridge was thought too expensive and a helicopter pad was ruled out on the grounds of noise and disturbance.

John Pitts produced *A Survey of Historical Documents relating to the Bindon Landslip* in 1974 and three years later he was well into proposals for research into *Mass Movement on Natural Slopes*. His potential research

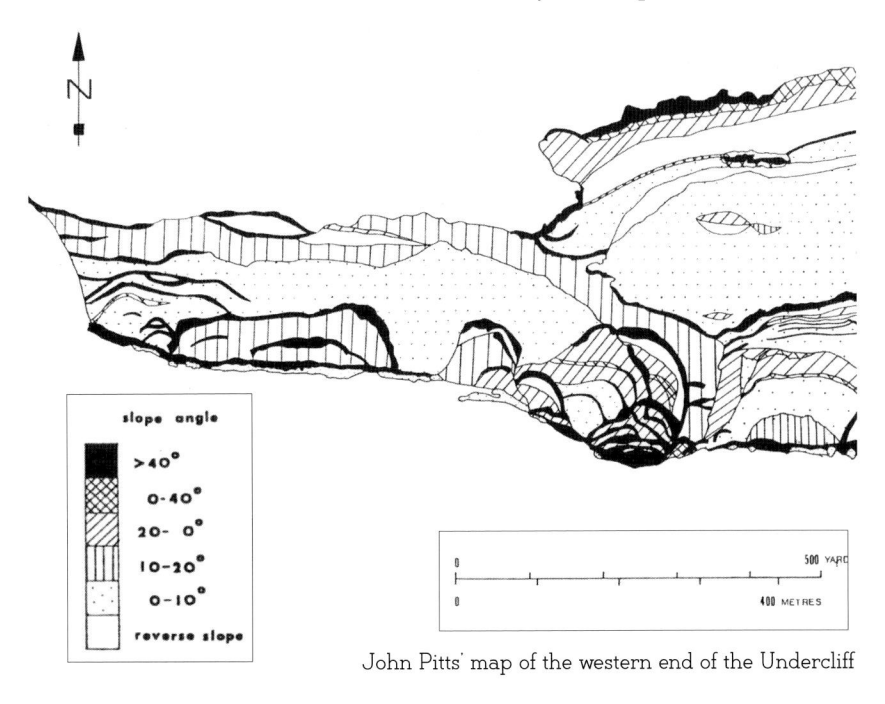

John Pitts' map of the western end of the Undercliff

Chapter 5

methods involved 'bore holes, trial pits and trenches' and his investigation of the 1839 event would involve drilling from the top of Goat Island through the Chalk to the Lias below. None of these methods were actually used but the complexity of what he eventually did can be appreciated by a look at one of his slope maps.

Tea at Landslip Cottage was very popular with visitors. The photos date from about 1934

When Tom Wallace was collecting information about life in and around Cliff Cottage, he visited Arthur and Elizabeth Critchard in April 1976, taking with him three photographs. One showed the cottage in 1937, the year of the sale of the Peek Estate, another had a lifebelt over the front door in 1928 and the third, from about 1905, included a Union Jack on a pole. Elizabeth's mother, Annie, had loved the cottage and eventually died there. Members of the family had lived in the cottage, with a few gaps, for over 50 years. They had always been dependent for their water on a shallow well slightly to the west of the house. When collecting it care was always needed to ensure the bottom sediment was not disturbed. Sometime in 1946 Annie had noticed and reported that the well water was murky. It turned out that two escaped German POWs who had set up camp nearby were responsible. Tom also heard from Elizabeth that the 'White Stone Path' along the cliff above had been gradually settling; its white stones acted as a guide to anyone walking home at night. He concluded that life in the cottage must have been hard with shopping needing a tough walk with a heavy basket.

Another source tells how a young Elizabeth used to meet her father at Combpyne station late on Saturday afternoons, after he had been shopping in Lyme, helping him carry groceries back into the Undercliff. She also fetched the cream for teas sold by Annie. Up to 150 of these might be sold on a Bank Holiday involving Annie in up to five trips to fetch more milk. The steep path up and down the cliff often had to be repaired because of minor slips but when it was fully cleared of brambles and scrub, it could be negotiated by a strong horse pulling a cart.

By 1978, Jim Kennard had become the chief warden and one of his reports tells of a wrecked trawler. *"During the night of the 1/2 December the Brixham steam trawler 'Fairway' of between 80 and 100 tons, went aground off the Undercliffs in the area known as the 'Slabs'. The trawler had lost her anchor in an incident off Brixham* and had drifted to Beer Head where the crew were taken off by Torbay Lifeboat – from there the gale had carried her across to the Undercliffs".

"When I visited the Reserve on the fifth, I found police with tracker dogs searching the undergrowth in the vicinity and I was informed by the sergeant in charge that a quantity of electronic equipment had been stolen from the trawler and that there was a possibility that this had been hidden in the Reserve".

"I was also informed by the police – a fact later confirmed by H M Coastguard – that because of the difficulties that would be involved in any salvage operations, the trawler is to be left to break up in its present position. It is well up on the beach and appears to be actually on the leased land of the Reserve... It is bound to cause a fair amount of interest and to draw a number of people to what is usually a little visited part of the Reserve. Whether, in fact, we can do anything about it, seems highly doubtful".

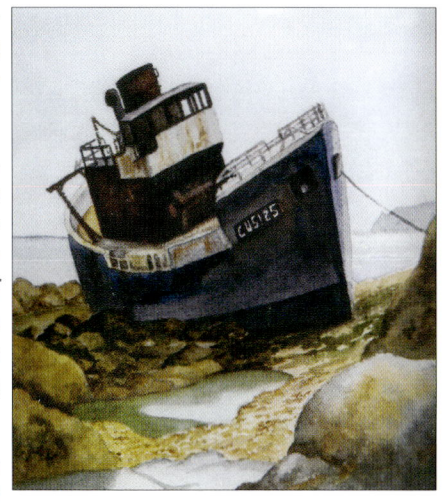

For many years there was a picture of the wreck in the bar of 'The Harbour' in Axmouth where she was described as being of 175 tons and as having broken in two before eventually being blown up.

The beached *Fairway* (from Margaret Pogson's diary of "My Devonshire Year").

The wrecked trawler and the remains of a digger which tried to salvage it are still rusting away on the beach.

A new era began in 1979 when Norman Barns became an Honorary Warden at a time of changing attitudes to the public as was indicated in an NCC Newsletter for the South West. It maintained that there had been a time when one of the first tasks associated with the establishment of a reserve was the erection of a barbed wire fence and keep out notices. Whereas reserves had been set up for science, they were now to be enjoyed by the public. The Chairman of the NCC strongly supported this view *"If the Council ever gets the reputation of being against public access, the NCC will positively be doing harm to conservation. Whenever reserves can cope with visitors, they should welcome them, elsewhere controlled access is the answer"*.

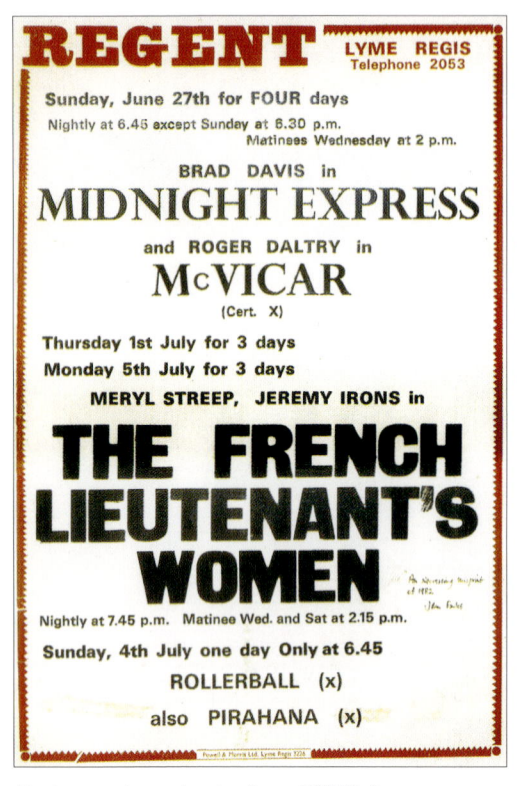

In April 1981, Muriel Arber, using slides of historic importance, gave a talk in Lyme Regis about the early history of local geology but her interest would soon be the film *The French Lieutenant's Woman*. It opened in Lyme on the same day as in Cambridge where Muriel lived. The film was just starting its ninth consecutive week in Cambridge when she wrote to John Fowles *"I am rather amused to look back on the telephone call I made to the cinema here to enquire when it would come here. They did not know but next day I tried again and was told it was coming this very Sunday, just for seven days, and it is still on nine weeks later! It is curious to feel that the Cobb and the Undercliff are almost permanently on now in St Andrew's Street here. I shall feel quite blank when the Undercliff is no longer there to be visited; these scenes are absolutely lovely"*.

"An interesting misprint from 1982". A comment from John Fowles who gave the poster to the Philpot Museum. This postcard version was sent to me by Sarah Fowles in March 2001.

After mentioning differences of colour on the covers of Robert's 1840 guide and wondering how Wanklyn would have rejoiced at various archives arriving at Lyme's Philpot Museum, she continued *"It is wonderful news that you are going to give a lecture on early fossiling at Lyme, at the Liverpool British Association meeting. I did not know about the cladistics battle at Liverpool. I knew about the controversy, of course. Beverley (Halstead) gave a lecture a year ago to the Geologists Association. Supposedly, about the first vertebrates, but it soon* became a diatribe about cladistics, with the BM (NH) and Marxism and so on involved".

"How very kind of you to offer to arrange for me to see the recent slipping of Underhill Farm in April. I should be most grateful. I have never been inside the grounds of the farm though we approached it by the fields that day some years ago when you took me down the cliff face and up to the fields – a never to be forgotten experience for me for which I have always been grateful. You showed me the maps of the farm and I have always remembered that the fields had fascinating names but did not remember what they were".

"Did you notice in the Times that at the moment of the earthquake during Mrs Thatcher's visit to Mexico she was having dinner with Crispin Tickell. Mary Anning's great grand nephew gave Mrs Thatcher dinner in Mexico in an earthquake; it sounds like a game of consequences".

John Fowles gave his talk in Liverpool and went to the essence of the Undercliffs, a World Heritage Site. "There are surely on occasions other values in the balance between the acquisition of new knowledge or university research. Public education is one. Another, if you will excuse me using such an unscientific word, is what I would call poetic – imparting a sense, however small, of the age and complexity of existence, both animate and inanimate, on this planet".

Another letter to John was in his capacity as Curator of the Philpot Museum. Norman Barns wrote about recent filming in the Undercliff after explaining his donation of 100 copies of his Undercliff guide to the Museum. "You may be interested to know that the BBC's 'The World About Us' team have filmed the whole length from a helicopter. Most of their close-up work on the ground was done in the spectacular area of Haven Cliff and 'The Elephant's Graveyard' through to 'Finger and Thumb'. Their aims are not so sensational, and their objectives are simpler than those of Television South West, but the treatment is more authentic and more serious. This particular item is 'Cliffs of Britain and Ireland' and in our vicinity they filmed at Cain's Folly and at Budleigh Salterton".

As Chief Warden, Jim Kennard, like all other National Nature Reserves Wardens, received a letter on 20 January 1984 telling of another television series featuring the natural history of the British Isles. He sent a note to Norman *"You may be hearing more of this".*

Chapter 5

Paintings by Elaine Franks from Geoffrey Young's Watching Wildlife.
From the top:
Sexton Beetle
Green Tiger Beetle
Bombardier

At the annual Programme Review a couple of months later it was thought that recent TV and film coverage had led to an increase in the use of the coast path. There was agreement that the Conservancy should oppose the proposed change to the route of the path as it was likely to cause disturbance. It had been proposed that it should go up Haven Cliff and along the cliff to meet the existing path below Bindon. More clearing and coppicing were needed all along the coast while Holm Oak should perhaps be controlled. No actual decision there! Fixed point photography was virtually impossible but regular records from reasonably secure points should be organised. Norman reported considerable movement at Pinhay Warren which was noticeable on a weekly or monthly basis.

In July 1985 someone other than Norman actually came to record some of the wildlife of the Reserve when M E Archer from York identified 83 species of aculeate wasps and bees. He maintained that the day was probably one of, if not the best, he had ever had.

Four years later there was even some Conservancy involvement when Alan Stubbs found high quality soft cliff habitat at the west and where he discovered a distinctive Spanish species of cranefly, *Helius hispanicus*, which was new to Britain. NCC's A O Chater and A P Fowles visited on 12 September looking for ants on the Plateau and finding *Leptothorax suberum* and four other species on their brief, two hour visit.

Another enthusiast for insects was short-sighted Elaine Franks who worked on her sketch book *The Undercliff* for two years before its publication in 1989. In the introduction she wrote that *"The easiest things to observe are those you can get close to – insects, for example. I have found that being very short-sighted has helped me to appreciate insects. When I was a child birds were those dark things that always seemed to be out of focus but a wasp busy at a pool of jam was a delight – the fascinating complexity of its structure and actions, the exquisite detail of its marking".*

She was, perhaps, particularly fond of beetles, illustrating a lovely Sexton Beetle, with a comment about its curved hind tibia, large and hungry Green Tiger Beetle and the Bombardier which deters predators with a jet of caustic liquid explosively ejected from its anus. A page of ladybirds, several of butterflies and, appropriately, just after my mention of Stubbs' Spanish Cranefly, she illustrated another Cranefly, *Tipula maxima*, hovering around a patch of mud before trampolining up and down on it, depositing eggs. However fond of insects she was, she painted plenty of other life, from Woodpeckers and Jays to Pygmy Shrews and a Smooth Newt, as well as Rock Roses and orchids and the tree lined track down to the old pumping station.

The 1990 Annual Review had some positive news with the appointment of Geoff Jones as the District Council's Rural Affairs Officer. Geoff was to become vital in bringing about the creation of Seaton Wetlands which 30 years later continues to develop. The Yarner team had, again, visited Goat Island for three days mowing and extending the chalk grassland areas which had improved so much in recent years. In addition, an access route to another pocket of grassland, Goat Island East, would enable future work to be done there. Woodroffe School had dug a new pond below Ware Cliffs and also improved the Chimney Rock footpath. Once again, Norman Barns shouldered the burden of educational work in his usual tireless manner.

The storm of 25 January 1991 must rank as the one event, which affected the Reserve more than any other, wrote the Chief Warden, Jim Kennard. *"The site is always subject to change by natural forces, but strong winds often pass it by without causing any marked damage. On this occasion, there was serious windthrow and shattering of standing trees throughout the whole Reserve, with the worst hit area appearing to be in the central section, where new vistas were opened up and many trees fell across the public footpath. So badly was the path blocked by 'barricades' of fallen timber, with some trunks precariously balanced on others, that the County Council closed the route pending clearance by contractors – a task for which NCC offered a financial contribution, in recognition of its responsibilities over the area".*

In 1992, a second Management Plan finally appeared having been prepared by the team of Jim Kennard (former Chief Warden), David Rogers (Chief Warden), Andrew King and Rob Wolton (Conservation Officers) as well as Norman. *"Naturalness is one of the principal attributes of the Reserve with much of the land untouched by the hand of man."* **Indeed, they comment,** *"it is one of the wildest and most unspoilt tracts of country in Southern Britain. This inherent naturalness is marred by the invasion on parts by exotic species, principally Holm Oak, which colonises recent landslips but also Sycamore, Buddleia and Pampas Grass. Where humans have been*

Table 2

Important Features of the NNR according to the Management Plan

	Importance		
	International	National	Regional
Geology			
Landslipping		Δ	
Jurassic stratigraphy	Δ		
Vertebrate palaeontology	Δ		
Rhaetian stratigraphy		Δ	
Biology			
Plant succession		Δ	
Naturalness		Δ	
Diversity			Δ
Calcareous Scrub / Grass			Δ
Calcareous Woodland, Ash		Δ	

active, the signs have largely disappeared under the developing woodland.

"With its softer more basic rocks, the Reserve would be expected to have a higher diversity of plant species than other coastal SSSI's in the South West. Apart from the seral sequence of habitats from newly exposed rock and soil through calcareous grassland and scrub to natural Ash woodland, there is rocky foreshore, wet flushes, seepages, ponds and small areas of plantation woodland.

"With regards to habitat rarity, the small pockets of chalk grassland that occur at the top of the cliffs are virtually the only representatives of this habitat in Devon, although it is much more common in Dorset. Otherwise, apart from climax Ash woodland (climax woodland is the final stage following natural succession), which is scarce in Devon, the other habitats are not uncommon in the context of protected areas either nationally or regionally, and this includes the limestone grassland.

On fragility they write that *"the Reserve could be considered fragile in terms of earth movements, that plateau grasslands are vulnerable to natural succession and that the rugged terrain and inaccessibility mean that little disturbance or damage from humans is likely.*

"The Reserve cannot be regarded as typical of any habitat type because no other site is totally comparable, its recorded history allows elucidation of the landslipping process but there are few published papers relating to the Reserve except those arising from research into geology and landslipping. Ecological links with nearby areas, vegetated cliff SSSI's and arable farmland, do occur but cannot be considered of much importance. There is potential value for woodland research under a non-intervention regime but, again, the hazards of the terrain limit the amount of research work and the educational and recreational use.

"The intrinsic appeal is high but limited through difficulties of access and lack of viewpoints. Important features of the Reserve are listed below."

Ten plant species of national importance were listed as important features together with 'insects' and the Lesser Horseshoe Bat. 'Insects' was a convenient broad-brush term used because no-one had much knowledge of what was there. More was known about the four regionally important species Peregrine Falcon, Nightingale, Lesser

Whitethroat and Dormouse. Maritime species dominated the list of nationally important plants which included Sea Kale, Rock Sea-Lavender, Sea Radish and Portland Spurge. Other species on the list were Dwarf Sedge, Early Gentian, Mountain (Pale) St John's-Wort, Nottingham Catchfly and Purple Gromwell. Although Corncockle was on the list, it had not been reported for years.

The plan continued with a selection of management constraints starting with the nature of the terrain. *"A second constraint is the lack of staff and finance. Local supervision has been maintained by dedicated Honorary and Volunteer Wardens while physical management has been carried out by English Nature staff based near Bovey Tracey or Wareham, both of which are more than an hour's drive away. Equipment and materials have been supplied from the same sources but there has been no specific financial allocation for the Reserve. Historically, the second constraint has doubtless been a product of the first, a reserve of this size and importance, which has existed for 35 years, would surely have been adequately staffed and financed by now had it been thought necessary. If full-time staff were appointed, it would be difficult to keep them busy"*.

It has long seemed to me that that was evading the issue! Even if the Undercliff had initially been seen primarily as a geological reserve, money and ambition had been lacking for 30 years and a mass of work that had been needed had never been done. Consequently, the whole Reserve had declined both in diversity and appeal. There were invertebrate groups that no-one had investigated since Tom Wallace and flowerless plant lists that had not changed for years. It was not only more recording that was needed, but a complete change of approach to management, if the Reserve was to justify its status. It was more than a part of the coastal geology where one unique event had happened years ago.

Tom Wallace had always championed the biological interest but now, having heard of the new Management Plan, he made contact with the Conservancy. In reply, he was told that things had altered a great deal since the time when Norman Moore was Regional Officer and that they were set to change a lot more in the next year or two. Many of the old names had gone and more were shortly to go. He was put in touch with Norman Barns and told that they might be able to find a spare Management Plan. When they did find one Tom was not happy with it as there was no reference to Hamish Archibald's original plan or to his own extensive contributions to it. His case was taken up by one of his former students, Sir John Lister-Kaye OBE, who had become Chairman of the North-West Region of Scottish Natural Heritage. Sir John thought it was no exaggeration to say that he had become a serious naturalist because of the Undercliff and the appreciation of scientific method and biological recording he had gained at school. He wrote to the Earl of Cranbrook, much involved in the English Nature hierarchy, to gain his support for Tom and to invite the Earl to Scotland to feed a visiting Pine Marten and to look at the Caley Pinewood in Glen Affric.

Despite there being five names at the head of the Management Plan, to which Wallace had objected, it would seem that David Rogers (Senior Site Manager 1990-1994) was largely responsible. When he was later interviewed about the plan by Albert Knott, he started by mentioning Laurie Pritchard's role as part-time Honorary Warden and the more active involvement of Norman Barns *"whose vast fund of knowledge of the site was always fascinating and sometimes useful"*. He then explained the organisation of the Reserve's management.

"Before my arrival, the Yarner team (one warden and three estate workers) had been developed by Robbie Roberts, based in Taunton, as a flying task force carrying out projects throughout the region. They were used to cut back vegetation around ponds and to control the grass and scrub on Goat Island once or twice a year to maintain the herb rich chalk grassland. The Purbeck team carried out the same function on the Plateau. This was done largely by hand and with clearing saws although on one occasion we dismantled an Allen Scythe, carried the pieces to Goat Island and reassembled it. That was so laborious that it was not repeated. I delegated the Undercliff management to Phil Page (Site Manager at Yarner Wood) as I had nine site managers to supervise and did not think I could give the Undercliff enough attention".

Later, in 1994, David became Senior Scientific Officer for the Maritime Team doing advisory and policy work on English coastal habitats. His team manager, Geoff Radley, saw soft cliffs as an important element of the vegetated sea cliff Special Areas of Conservation and between them they revised the criteria for these sites,

Before the establishment of the NNR, rabbits had done the mowing of the more extensive Plateau grassland.

Chapter 5

considering the Undercliff as among the best soft cliff sites in Britain because of its complex and well-studied geology, its active geological processes and varied vegetation. They also supported the submission for the coast to become a World Heritage Site.

Around the time of this World Heritage designation, I interviewed Phil Page at the English Nature headquarters in Exeter. Although he did not know much about the early days, he did know about things since 1986 when he had started working in the region. He first visited the Reserve in March 1987 coming with Chief Warden, Jim Kennard, who had been managing it with Norman. Later, Phil and other staff from Yarner met up with three NCC workers and two volunteers from Dorset for a half day on the Plateau before more cutting and raking on Goat Island in the afternoon. His recollection was that both areas were fairly overgrown, the clearings were smaller than at the time of the interview, and that Jim was considering abandoning the grasslands. Phil disagreed and thought they could be managed more effectively without too much cost.

In August 1981 scrub was developing on the Plateau

"What we needed to do was to stay locally, take tools down and leave them there. The path down was not clear at the time but after some cutting back, we made good progress". Phil thought this work should not be wasted so he arranged a three day visit for spring 1988 and gradually they began to cut back scrub as well. *"When the Nature Conservancy became English Nature in 1991, David Rogers took over from Jim Kennard and English Nature started mowing the Plateau as well, but that was too much, and contractors were called in. That was possible because the Reserve started getting money, work expanded and to say otherwise, as Norman did, was untrue!*

"Historically, under the Nature Conservancy the bulk of any money from the mid-1970s went to the Dorset and Lizard Peninsula heathlands but when it became English Nature and we were reorganised as Devon

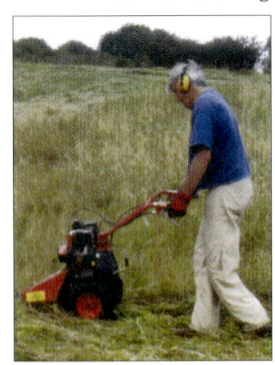

Phil Page mowing on Goat Island in about 2000

and Cornwall, we got a bit more money and when we became just a Devon team, the situation improved again. Norman's story is contrary to my experience: I have seen photographs of him there in the 1970s surrounded by tall grass. Really that is what I wanted to say, I wanted to put the record straight".

When I asked him to say something about the organisation of the Devon team, Phil told of *"a team manager and his deputy, six conservation officers and three assistants, myself as site manager and Albert Knott as my assistant. There was also an estate worker who sometimes had an assistant. We had amalgamated three Dartmoor reserves, Yarner Wood, Trendlebeer Down and the Bovey Valley woodlands to form East Devon Woods"*. Asked about the CROW Act, he did not think it would have much effect on the National Park, but its bureaucracy and extra powers could lead to an extra workload elsewhere with the drawing up of mini management plans for every SSSI and quite a bit of extra work as there is with any new system.

The Plateau after its annual mow

Back in the Undercliff, or very close to it, there had been two proposals for field study centres. One suggestion was for the conversion of redundant farm buildings at Ware, 100m beyond the NNR boundary, and another was for the conversion of Cliff Cottage, also just outside the Reserve, for the benefit of Allhallows students from the school above the cottage. Although ready to support controlled educational use of the Reserve, David Rogers considered that English Nature could not get involved in local politics but could support controlled educational use of the Reserve.

Neither venture materialised and nor did an Undercliffs exhibition planned for summer 1995. Allhallows was, again, the focus of activity with teacher Keith Moore to co-ordinate and develop the geology theme, Norman to do the same for natural history and Peter Jackson to deal with aspects of social history. Others would collect information about Lyme's cement works and railway, about grazing on the Undercliff, collecting photographs and perhaps restoring the sheep-wash.

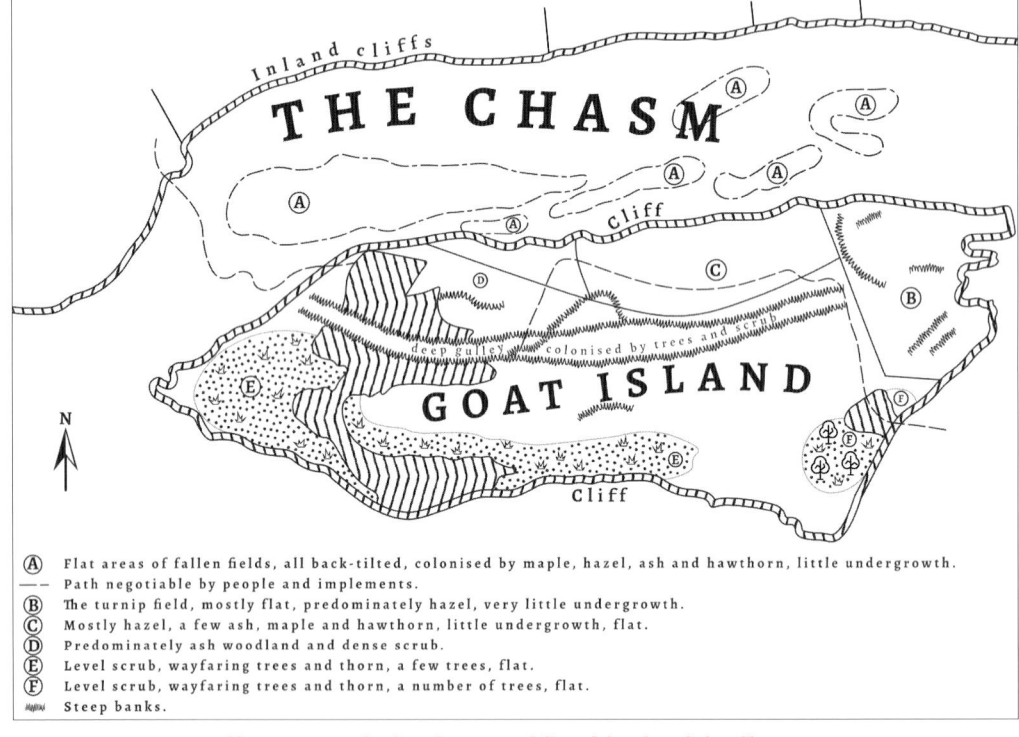

Ⓐ   Flat areas of fallen fields, all back-tilted, colonised by maple, hazel, ash and hawthorn, little undergrowth.
---   Path negotiable by people and implements.
Ⓑ   The turnip field, mostly flat, predominately hazel, very little undergrowth.
Ⓒ   Mostly hazel, a few ash, maple and hawthorn, little undergrowth, flat.
Ⓓ   Predominately ash woodland and dense scrub.
Ⓔ   Level scrub, wayfaring trees and thorn, a few trees, flat.
Ⓕ   Level scrub, wayfaring trees and thorn, a number of trees, flat.
    Steep banks.

Vegetation and other features of Goat Island and the Chasm.
After Norman Barns (1985?)

CHAPTER SIX

Norman Barns, Management Plans and Early Surveys

In 1952, Norman Barns, a lifetime enthusiast for the Undercliff, received a note from the Headmaster of Allhallows *"by all means make your way to the landslip through our grounds"*. When I interviewed Norman in 2001, he recalled his first walk among the cliffs in 1927 and remembered his struggles when the path was not discernible *"boy scouts would do the clearing, but great chunks would fall down, and my little feet were not good at it"*.

More than 50 years later he was the authority on the area in his role as the Nature Conservancy's Warden. His love of the Undercliff was evident in an article for the Conservancy's Newsletter of 7 December 1980. He wrote at length, 3,000 words, so only a shortened version follows:

"Now 9.30: sandwiches made, vacuum flask filled, and I am ready for off. Secateurs, pruning saw, pocketknife, hank of hemp sash cord, binoculars, walking stick, two wardening authority cards, armband, badge and a few copies of the NCC Reserve leaflet". He drove to Barn Close Lane, walked across three fields, admired the view to Beer and listened to the Skylarks before dropping down into the Reserve. He was soon standing *"among the rocks between Goat Island and the sea, looking up at the chalk cliff. It is a long time since I saw a Peregrine there. A pair of Herring Gulls are busy on a ledge. They are not nesting but have each taken a shore crab up there for their elevenses. On a cliff edge Hawthorn, a Raven looks down on them hopefully. He looks ridiculously large on so small a bush. Goat Island is no longer a real island but a ten acre plateau surrounded by huge rockfalls that are like petrified waves on a giant sea"*.

Below Dowlands *"the path is about two feet wide with Dog's Mercury verges. Here there are 40 foot high Hazels. It is very shady here as all the trees hang heavy with ivy. Here is one that has toppled over*

Chapter 6

– and across the path! There is already a worn detour around it. Blade-guard off the saw and set to work. Three boughs about six inches across are first to be sawn through. After a quarter of an hour the job is done, so on through the Hazel grove. Up the steep slope and down the next gully which is like an Ash tree graveyard. Trees litter the place while moribund specimens wince under the weight of Ivy, which is slaughtering hundreds of trees by its sheer weight, long before they reach maturity. It is a sad sight all over the Reserve. There are vast areas where once the chalk banks were bright with Rockrose, Harebell and Chicory and now Ivy covers the ground. Its evergreen foliage keeps light away from the ground plants. On through the first real mud and nearing Whitlands. On the path I meet a weary party seated by the path. 'We thought we would try it but we are not going any further'. 'I thought you Scots were tough' I say cheerily. 'Aye, we are – in Scotland!' I leave them and move on under taller trees.

"I am surprised when a couple of Roe Deer come near. They are a sandy colour at this time of year and they look married, or at least they have a serious relationship going, as they walk along, nuzzling each other's necks. They stop and look unconcernedly at me. I ask them quietly how they are. If I were to get up or move, they would be dashing away, leaping over low obstacles, but quiet talking never seems to bother them. I wish I had remembered some chocolate. They love that. Eventually they go but the place is not long empty. A fox comes speculating. An old fox I guess because he does not come too close but sits down a short way off and looks at me".

Having reached Underhill Farm, he turns and retraces his route back to Pinhay and then Whitlands. There, meeting a couple who have walked from Seaton, he says his usual piece about the Nature Conservancy only to receive the comment *"I think they do wonderful work, I have been a member for five years"*. Norman reckoned that people could not conceive that a Government Department would be concerned with nature conservation. His next encounter is with a family coming up from the beach near the 'Slabs'. He talks to them about getting a permit if they wish to leave the one definitive path. On his way along the beach towards the mouth of the Axe he admires Yellow-horned Poppies and Sea Spurrey and at the Harbour he is surprised and pleased to find the Reserve signs undamaged.

Norman then clambered up the previous winter's cliff fall to a wild, wide shelf below the Chalk and Sandstone at the top of Haven Cliff. *"A dull coloured Adder basks in a hollow. It is a good area for land snail*

Norman Barns, Management Plans and Early Surveys

"A wide, wild shelf below the Chalk and Sandstone (Nature Conservancy 1981) with no sign of a path.

NCC had recorded a path in 1964

"See where that river goes into the sea, that's called 'the marf'".

103

Chapter 6

species. The grass is covered with empty mussel shells and crab carapaces, remains of gull lunches. Six o'clock. I will sit here for a while". Returning to the beach, he meets a middle aged camper and his wife. They are enthusiastic. *"Grand, ain't it, all this. It's all natural and it's all clean. People take trouble too. It's not like Birmingham"*. He turns to his wife *"See where that river goes into the sea, that's called the marf. There's a lot of rivers ain't there. There's the Dart and the Teign and they've all got marfs"*. Norman, enthusiastic as ever, also taught creative writing.

During his walk he had met 40 foot Hazels and blamed Ivy for the Ash tree graveyard. In the first Management Plan, Hamish Archibald had mentioned a Hazel with a 38 inch girth and rather surprisingly wrote that *"It will be most interesting to observe what species (will) succeed Ash on this site"*. He also observed that seed was available from a wide range of trees and shrubs and that many of them had already proved their ability to germinate and to grow well.

Appendix three of that Plan is a list of woody species found in the Undercliff. Of larger mature trees only Ash, Field Maple, two Birches, Silver and Downy, and perhaps Sycamore and the three scattered Hornbeams were considered native. The other 24, including three Oak species, were thought to have been planted. Among the 35 small trees and larger shrubs he listed 19 as occurring naturally and 16 as planted while among the 23 smaller under-shrubs 13 were likely to be native. Thus, out of 88 species, 50 were thought to have been planted. Of these, eight were conifers. Plenty of the others would not long survive the competition as larger trees matured and, like the spreading Holm Oak, cast increasing shade.

Aerial photographs of the land between Humble Point and the old coast path from Whitlands towards the sea showed extensive open ground in 1950. By 1957, by which time the NNR had been designated, there were scattered Holm Oak individuals but even in 1990 much was still treeless in the area through which the coast path passed on its way towards the pumping station. By 2003 the individual trees visible from the air had been replaced by extensive patches of Holm Oak.

I feel, without much evidence, that Hamish was more familiar with the east and central parts of the Reserve and that he looked carefully around old habitations for signs of fruit bushes. Four Currants and six species of Prunus soon disappeared. A Medlar survives near Ames's Wall near Ravine Pond.

Table 3

The status of larger trees in 1955 with the circumference of some at that time, and again in 2021.

Key to the DAFOR scale: D=Dominant, A=Abundant, F=Frequent, O=Occasional, R=Rare, L = Locally.

Species		Age / Maturity			Max Circumference (cm)	
Scientific Name	English Name	Planted	Native	20+ Native	1955	2021
Acer campestre	Field Maple		O	O/F		297
Acer pseudoplatanus	Sycamore	LF			98	295
Aesculus hippocastanum	Horse Chestnut	R (3)			65	(340)
Alnus glutinosa	Alder			O		75
Betula pubescens	Downy Birch		R	LF		
Betula pendula	Silver Birch		R	LF	60	135
Carpinus betulus	Hornbeam	R (3)	?		48	230
Castanea sativa	Sweet Chestnut	R				460
Fagus sylvatica	Beech	LF			175	370
Fraxinus excelsior	Ash	?	LF	LD	176	338
Juglans regia	Walnut	R (3)			57	
Larix decidua	Larch	R		LF	44	
Picea abies	Norway Spruce			Planted LF	48	220
Pinus sylvestris	Scots Pine	O/R				
Prunus avium	Wild Cherry	R		R	50	
Pyrus communis	Cultivated Pear	R				
Quercus cerris	Turkey Oak	LF/O		O	75	420
Quercus ilex	Holm Oak	LF		LF	90	480

105

Species		Age / Maturity			Max Circumference (cm)	
Scientific Name	English Name	Planted	Native	20+ Native	1955	2021
Quercus robur	Pedunculate Oak	R			65	280
Taxus baccata	Yew					280
Thuja plicata	Western Red Cedar	LF			69	
Tilia x vulgaris	Hybrid Lime	R			125	430
Ulmus carpinifolia	Smooth Elm	R		O	69	150
Ulmus glabra	Wych Elm	R		R		

Among the small trees and larger shrubs, Hazel was, and is, locally abundant. Seven species are described as locally frequent while Hawthorn and Elder are rated occasional. The seven were Blackthorn, Buddleia, Grey Willow, Spindle, Wayfaring Tree and *Rhododendron ponticum*, which has now almost been eliminated except in the 'Lost Garden'. The battle against Buddleia continues, particularly on Haven Cliff.

35 years later Field Maple had become widespread as indicated by Norman's summary of plants along the coast path and on the shore. Hawthorn was now more common than Hamish had found it. Turkey and Pedunculate Oaks occurred at Ware and there were two of the latter at Dowlands. A Hornbeam near the pumping station was also on his list. Dog's Mercury, Bugle and Wood Sanicle were common along the path together with Pendulous and Wood Sedges. Pendulous Sedge has become even more common with the deer leaving it uneaten. Perforate St-John's-wort was found below Goat Island, above Charton Bay, and near West Cliff Cottage, with Navelwort near the Pinhay fault, and Meadowsweet towards the east end. Since Norman first recorded the plants Cotton-grass, Green-winged Orchids, Heather and Wood Violet are among some thirty species, including a number from along the coast, which have disappeared.

The shade of the growing tree cover, the many little water courses paddling their way towards the sea and the damp Devon weather are ideal for ferns. Back in the days of the Victorian fern mania, the

Norman Barns, Management Plans and Early Surveys

Reverend Zachary James Edwards, Rector of Combpyne, wrote a miniature classic *The Ferns of the Axe and its Tributaries* (1862). He justified his book by *"venturing to remark, in this utilitarian age, that the student of natural history is pursuing no useless occupation, when he devotes himself to the contemplation of the beauteous works of God."* Later, in a different context, *"May we not learn, even from nature's works, to be steadfast in purpose and firm in conduct not vacillating on every trifling occasion."*

In a recent article about the Reverend Edward Peek (in *Museum Friend* July 2019) Max Hebditch describes how Zachary Edwards praised the Reverend, brother of Sir Henry, for his work in the restoration of the Guildhall. At that time, 1889, Edwards was Mayor of Lyme but he was succeded in 1890 by Thomas Philpot who soon hosted a historical exhibition and dinner in the Guildhall. After the dinner Edwards, as local historian, Latin scholar and intellectual, gave a talk. Shortly afterwards Philpot, Edwards or Peek suggested that there ought to be a permanent museum in Lyme.

In his fern book Edwards had mentioned two fern species from the Undercliff. He had found Sea Spleenwort on rocks besides Charton Bay and Wall-rue on Ware Cliff. They are not there today but a few Black Spleenwort can be found in Humble Glades and a single one may still survive on the top of a pinnacle at the east end of the Chasm. Too steep for me!

In much of the reserve Hart's Tongue is dominant often sharing this status with Wild Garlic as it does below Whitlands and by the path below Chimney Rock. Polypody is another distinctive species growing on many of the near-horizontal tree branches. In May 2021 the ferns looked very much in need of rain but they didn't need to wait long.

Edwards' complex Plate 1 (right) shows ten species, five of which can be found in the Undercliff today. The Polypody (3), Soft Shield (4), Male Fern (7), Broad Buckler (8) and Hard Fern (18). His second illustration includes Lady Fern and Black Spleenwort among its nine species.

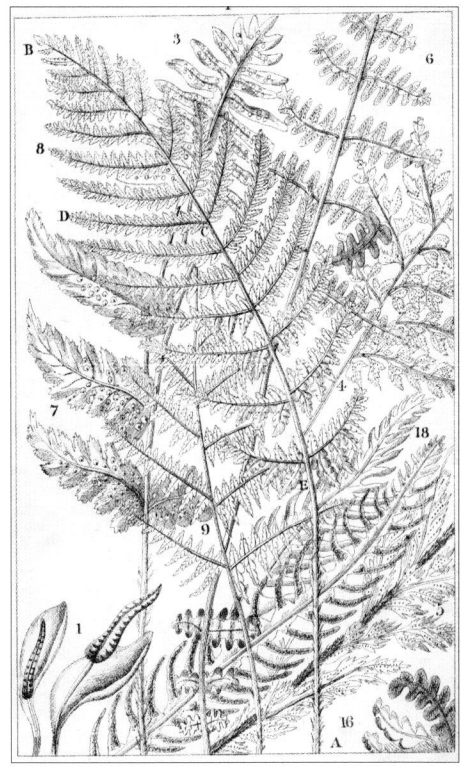

Chapter 6

A and G Lister produced a handwritten list of 184 moss varieties found in the neighbourhood of Lyme between 1884 and 1905. They only include details of the first record of each species found, thirty-two of which came from the Undercliff. In 1961, Tom Wallace produced lists of 16 liverworts and 75 mosses crediting Dr M C F Proctor of the University of Exeter for most of the records. *"On his several visits to the Landslip with students he has prepared lists of his collections and has very kindly identified or confirmed other material collected on the Landslip during field studies by members of the school. The Reserve with its humid climate, broken terrain, calcareous rocks and undisturbed dense vegetation, should contain a rich bryophyte flora"*.

In a letter in 2001, Mark Pool, commenting on the Wallace/Proctor summary wrote that *"it contains one or two unusual species although none are nationally rare or endangered.* Scorpiurum circinatum *is a Mediterranean-Atlantic moss which reaches its northern limit in the British Isles. The liverwort* Cololejeunea minutissima, *frequent in the Reserve, has a rather similar distribution.* Isothecium (now Eurhynchium) stridulum *is a very local species nationally"*. As described in chapter seven, Mark and Roy Jefferies spent a profitable two days in the Reserve in 2002 finding 100 moss species and 21 liverworts including *Frullania tamarisci* 'apparently very rare' on dying Birch at c.309903. Mark was back for the 2011 Bioblitz when he added ten moss species bringing the Reserve list to 137.

In the Cox survey of 1995, the Culverhole flushes were the most productive areas for moss species, as listed in Table 4.

The sometimes unchanging nature of nature was demonstrated when a curious small form of *Scorpiurium circinatum* was found near the base of a Holm Oak at Pinhay where Professor Richards had found it 40 years previously.

Since Jeff Benn's visit back in October 2001, botanist, David Allen, has been most alert to any fungi of interest. Quite a high proportion of the records he summarised in 2015 date back to an all-day foray in 2001.

Of the 24 Agarics mentioned in David's summary, 12 had been found in 2001, as had 2 out of 5 Boletes, 3 of 9 Polypores and 5 out of 7 Ascomycetes. Some of the Boletes have been among the most exciting finds. *Boletus satanus*, the Devil's Bolete, was first found by Phil Parr in October 2007 and his identification was confirmed two years later when he and David Allen found it again in July. At another site, under Holm

Table 4

Mosses found by W and J Cox in Culverhole on 23 July 1995. (Some of these names have recently changed.)

Brachythecium rivulare	O (F)	*Fissidens adianthoides*	F (O)
Bryum pseudotriquetrum	O	*Gymnostomum recurvirostrum*	R
Calliergonella cuspidata	O (F)	*Neckera complanata*	F
Campylium stellatum	O (F)	*Pseudoscleropodium purum*	R
Cephaloziella sp	O (F)	*Rhynchostegiella tenella*	O (F)
Ceratodon purpureus	R	*Seligeria sp*	R
Cratoneuron commutatum	R	*Thamnobryum alopecuroides*	A (F)
Eurhynchium praelongum	O	*Tortella flavovirens*	R (O)
E. speciosum	O	*Trichostomum crispulum*	R
E. striatum	O		

Oak, at the eastern end of Charton Bay, SY 303 900, David found ten in August 2011. Two other Boletes, *luridus* and *satanoides*, were recorded, the first on Goat Island in August 2002 and 2011, and the other, west of Whitlands, in August 2011.

Among Gastromycetes *"the rare, extraordinary, scarlet and decidedly malodorous Clathrus ruber"* as Wallace described it, was found by some of his students three times in one day in June 1958 and others found it every year for at least the next six. I found it on Haven Cliff in 2010 and several times since. It has also turned up on the slope of Goat Island east. David had an Earth Star, *Geastrum sp*, near Chapel Rock in September 2013 and Phil's reports have *Pisoldius arhizus* on soft cliffs above Charton Bay in August 2008 and two good records the next year; a Warted Amanita on the beach road below Rousdon and a Spectacular Rustgill near Humble Road. Six bracket fungi or Polypores are on the list including *Grifola frondosa*, mentioned earlier when Jeff Benn recorded it in the Undercliff. It is described in one popular field guide as *"Densely tufted, branched, spatulate to narrowly fan shaped, imbricated, greyish brown caps"*. David also found, and had the identification confirmed, three tooth fungi, all edible, and all previously referred to the single genus *Hydnum*. Wallace and others mention two familiar species, the well named Stinkhorn, also known suitably as *Phallus impudicus*, and the black crusty Cramp Balls (*Daldinia concentrica*) which some put in their beds to sort out the cramp.

Chapter 6

The Agarics found by Jeff and David, recording for English Nature in 2001, were mentioned earlier. Those found at other times are listed in Table 5.

For many years I loved the somewhat chaotic rocky oak woods in the Vale of Ffestiniog and along Afon Artro and Nantcol. More recently, I have got to know something of the moss and lichen havens among the Atlantic woodlands of Knoydart and Ardnamurchan, known to some as the Celtic rainforest. The Undercliff, too, is good at being somewhat chaotic but its mosses and lichens rarely play the dominant role that they can do when age has shaped each standing tree and eventually brought about the disintegration of the fallen ones. Above the River Coly, a few miles away, is neglected woodland on steep slopes beside a public footpath which is usually a stream. Having been measuring the girth of Undercliff trees, I have become more aware of the extraordinary size of some trees in this wood and in the hedgerows around Colyton. The town is set in an ancient landscape which stands in contrast to the no doubt efficient, but wildlife deprived and hedge depleted farmland, above the Undercliff.

Sixty years ago, as a potential doctor at University, I memorised a passage from an unremembered source. *"The inadequate conception that science is measurement and concerns itself with nothing but the metrical has become a thought cramping obsession, and the*

Table 5
Some Agarics from the Undercliff. The Milkcap is nationally Occasional.

Agaricus lanipes	Goat Island – October 2006
Agaricus xanthodermus	Whitlands – October 2013
Clitocybe fragrans	Rousdon – October 1998
Clitopilus hobsonii	Rousdon and Whitlands – November 2011
Collybia confluens	Rousdon – October 1998
Inocybe bongardii	Rousdon – October 1998
Lactarius acerrionus	The Milkcap: near Humble Pond – September 2009
Pleurotus pulmonarius	On Ash at Rousdon – October 1988 and on Beech, Whitlands
Russula mairei	Under Beach, Whitlands – October 2013
Russula sororia	Whitlands – August 2011
Tubaria conspersa	June 1999

more nearly a scientific paper approximates to a long and bloodless caravan of equations plodding across the desert pages of some obscure journal, the more quintessentially scientific it is supposed to be though no-one can tell, and few are interested to ask, whither in the field of human knowledge the caravan is bound".

Some modern books on natural history, determined to be scientific, have lost the spontaneity and imagination of older ones but Oliver Gilbert's *Lichens* in the *New Naturalist* series retains the feeling of joy at discovery and wonder at gaining greater understanding. The pleasure in discovery is evident when the author describes spending thirty minutes or so collecting invertebrates off lichen covered trees. His lichens, leafy and crustose, were concentrated on the weather-exposed surface of the tree trunk, with the other side either bare or with its bark covered with powdery algae. Bark-lice were common, feeding on the algae, while moth caterpillars fed on the lichens. Carnivores were common with plant bugs, harvestmen, Lacewing larvae and spiders using the cover for refuge and protection from desiccation during the

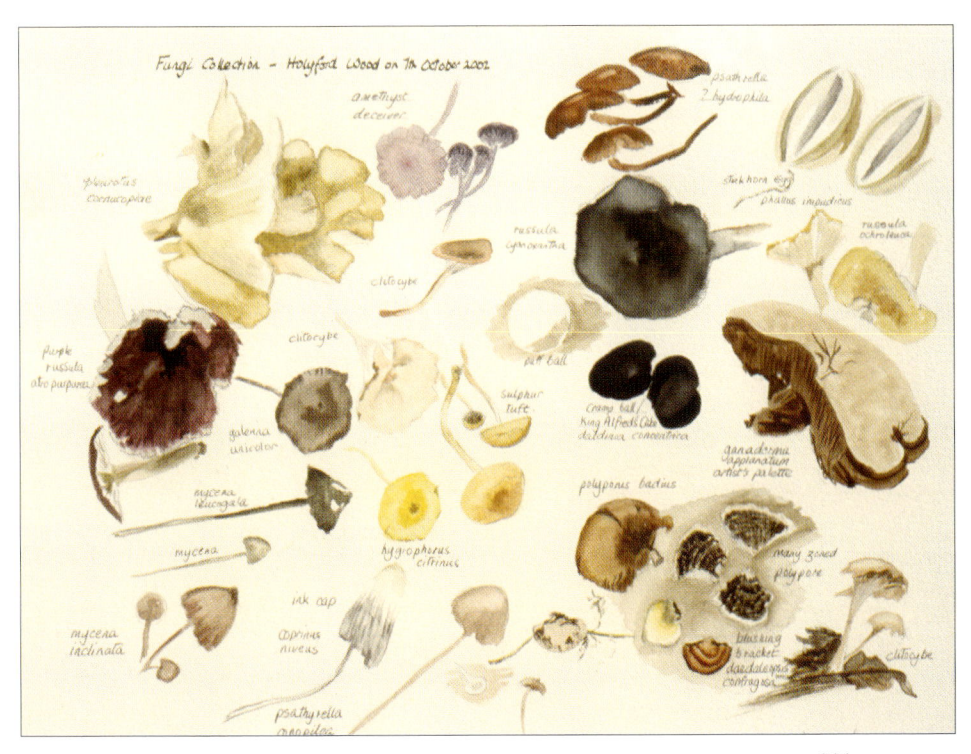

A collection of fungi from Holyford Woods, 2002.

Chapter 6

day before hunting over the tree at night. The small number of nocturnal moths and caddis flies were so well camouflaged that Gilbert felt he may well have missed many of them. As with the food-web surrounding woodlice described by Stephen Sutton in Chapter 7 complexity has close links to stability.

Gilbert's second chapter begins by asking what is a lichen and answers that in the old days, when life was simple, every school child knew that a lichen was a symbiotic association between a fungus and an alga. I rather doubt if they did, but in any case, as blue green algae, which are not algae, can be one of the partners, the definition changed to 'an

Val Baker's version of Oliver Gilbert's observations of life on a tree trunk where lichens flourish.

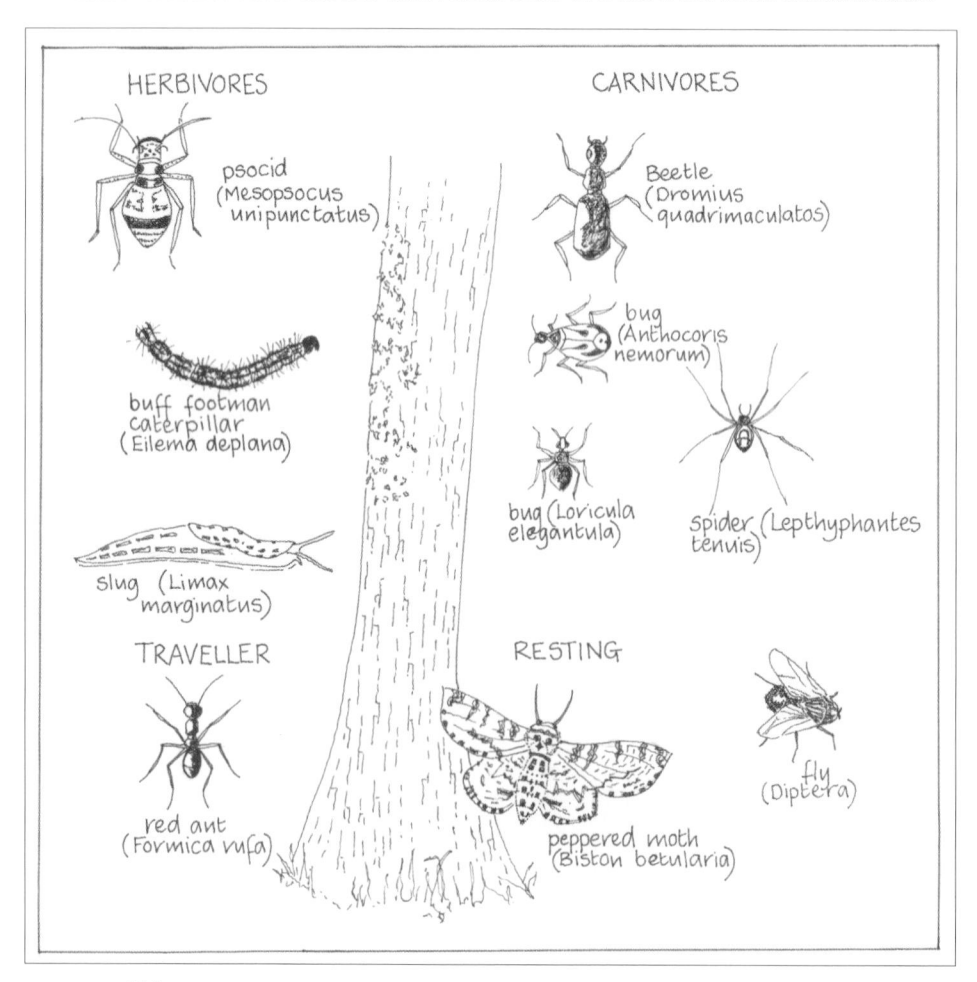

HERBIVORES
- psocid (Mesopsocus unipunctatus)
- buff footman caterpillar (Eilema deplana)
- slug (Limax marginatus)

CARNIVORES
- Beetle (Dromius quadrimaculatos)
- bug (Anthocoris nemorum)
- bug (Loricula elegantula)
- spider (Lepthyphantes tenuis)

TRAVELLER
- red ant (Formica rufa)

RESTING
- peppered moth (Biston betularia)
- fly (Diptera)

association of a fungus and a photosynthetic symbiont'. Blue green algae became Cyanobacteria.

However, some brown seaweeds invariably support fungi which do no harm, but they are not considered lichens. David Hawksworth, much involved with attempts at a definition, suggested that a lichen was an organism studied by a lichenologist but more usefully, in his *Dictionary of Fungi*, two definitions are given. One is simple *"A lichen is a self-supporting association of a fungus (mycobiont) and an alga or cyanobacterium (photobiont)"*. A more complex and precise definition is that *"A lichen is an ecologically obligate, stable mutualism between an exhabitant fungal partner and an inhabitant population of extracellularly located unicellular or filamentous algal or cyanobacterial cells"*. As Gilbert says *"Take your choice"*. Perhaps they are not so much a taxonomic group as a fungal lifestyle.

On 6 June 1972, Dr Hawksworth visited the Reserve on behalf of the NCC. He concentrated his search on the area between Bindon cliffs and Whitlands, investigating woodland and sandstone dwelling forms. He found 89 species, just about what he expected for Devon, with the number limited by the scarcity of mature deciduous trees and the absence of cliffs formed of hard siliceous rocks. Some of the communities on the Reserve contained species which he had found at other Devon sites, Berry Head, Chudleigh and Buckfastleigh, with limestone outcrops, but other species were known from few other sites in the county. These included *Arthropyrenia conoidea, Caloplaca ochracea, Protoblastenia monticola* and *Squamarina crassa*.

Among the stone loving species, he found an interesting community on a group of massive limestone blocks north-west of Humble Point. Barbara Benfield, photographed in Gilbert's book, thinks that *Squamarina crassa* was the most interesting species found.

Siliceous rocks at Humble Point and elsewhere *"Appeared to have been scrubbed clean and this appearance is typical of damage by oil pollutions. Lumps of old oil were noted among the pebbles on the storm ridges of beaches on the east side of Charton Bay. No species of Lichina or Ramalina were found in this habitat"*. Twenty years later, Ambios Environmental Consultants, investigating the coast, found *Lichina pygmaea* thriving on vertical surfaces in sunny situations along the lower littoral fringe. Patches were found at Pinhay west where *Verrucaria maura* and grey unidentified lichens were found on the supralittoral fringe. Some of the *Verrucaria* had survived the oil but were found, as poorly developed scraps, in areas protected from it.

Chapter 6

Lichina pygmaea, wrote Gilbert, requires more regular submersion than any other macro lichen and was for a long time regarded as a diminutive brown alga which went under the name of *Fucus pygmaea* so that in the older floras it must be sought in the algal section. Barbara Benfield was to add a number of species to Hawksworth's list when she recorded 84 species on visits in 1997 and 1999 in both of which she started from Ware and worked her way towards Pinhay.

Two final oddities from the lichen world, one involving Long-tailed Tits and the other a Herbarium in Nottingham. When making their complex nests, the tits are selective in their choice of lichens used for camouflage. *Parmelia perlata, P. sulcata* and *Physcia tenella*, all common in the Undercliff, are chosen by the tits, in preference to others. Examination of 17 nests from different parts of the country showed that the birds used a 'Velcro' principle, hooking the rhizinae on the underside of a lichen fragment to the structural core of the nest camouflaging it with grey green colours.

Long-tailed tits are selective in their choice of lichens.

The second oddity concerns an addendum Dr Hawksworth included in his report. He had come across a 19th century record *"presumably from the area now in the Reserve"*. It involved the lichen *Pseudocyphellaria aurata* in E M Holme's herbarium at Wollaton Hall, Nottingham with a duplicate in the British Museum (Natural History), It was labelled 'Undercliffs, Lyme Regis, on apple trees, A Lister Esq., the only British locality'. This is not mentioned in either Holmes's or Parfitt's accounts of the lichens of Devonshire and was almost certainly discovered in the Undercliff between 1883 and 1886 as in the latter year the new name *Stricta aurata sub-glaucescens* was published and the Lister lichen is the type specimen of this form.

This chapter began with Norman struggling through the trees in 1927 and getting permission to 'make his way to the Landslip' in 1952. In 1979, he had become the official warden and in 2001, when I interviewed him, he was enthusiastic about 1980s co-operation but less happy about recent events. *"In the 1980s, people from Slepe Farm, near Wareham, came to do the Plateau work, Dorset Countryside Service created most of the paths and office staff from Somerset came for picnic weekends when they provided an enormous amount of labour. At that time, the Reserve was famous for wild flowers and butterflies and there were lots of visitors but not many geologists. We took great pride in the glades, their design and creation, helped by Allhallows pupils"*.

Asked whether people actually knew much about the Reserve's wildlife, Norman sadly said *"I think that is history now, the wild flowers are no longer there. I think we had good knowledge because I spent most of my time there, botanising, searching every part regularly and finding Harebells, for instance, despite their leaves and stalks looking just like grass and with flowers that do not last long. I think we knew a good deal about what was where; well over 500 plant species, Tom's fungi lists, and papers produced by all sorts of people with special interests"*.

Norman went on to describe some of the major changes of which the encroachment of Holm Oak was the worst. *"At one time, all over the Plateau, you found a seedling every square metre but good mowing practice, well timed, gradually reduced them. Even with the mowing there were huge reductions in Autumn Gentian numbers. We used to count 2,000, we got fed up with counting. (In recent years there have been estimates of 30,000 even if in other years there are only a few). There were many more of the stiff petalled Carline Thistle. Lost glades, encroachment of non-indigenous species, less light everywhere, so violets decreased and so, of course, did the fritillary butterflies"*.

Norman plants a tree in Combpyne churchyard

When asked about his involvement in the second Management Plan and whether it had been implemented, he said that it had been *"because it is easy to implement a non-intervention policy. We didn't like doing the Plan as it meant work stopped. There has never been a reserve of comparable size with so little money spent. I was virtually the only cost and had worked voluntarily for many years before receiving an honorarium. It was an excuse for neglect, for abandonment really. When it was said that this or that must be monitored no-one was appointed to do the monitoring. I had a man, he came down to collect evidence for a book on 'Invertebrates of Soft Rock Areas of the Coast' and having arrived*

Norman gardening in Combpyne

Chapter 6

at 2.30pm, he left at 3.45pm". Asked if he felt frustrated, he replied that *"it was terrible, but they pay me and that is all the time they give me"*!

Norman thought that they considered it as a geological reserve and that geology does not need management. I asked whether Tom Wallace had had harmonious relations with the professionals of English Nature and the reply was that, as previously mentioned, he resented the lack of reference to Hamish Archibald's first plan and the contributions that he, Tom, had made to it. Norman also disliked what he saw as changed attitudes on the part of the professionals. He felt that the contributions of volunteers were no longer appreciated in the same way as when Jim Kennard was Warden. *"It was wonderful then when office staff and all sorts of people used to come and work on the Reserve"*. Adrian Bown, the other voluntary Warden, also felt disillusioned but continued to work effectively having, according to Norman, got the Plateau *"looking really good, like a fairway."*

There were many more of the stiff petalled Carline Thistle

Norman was President of the local Conservation Society for many years and after retiring from teaching was spending up to 200 days a year in the Reserve. He also led some 40 walks around the Axe catchment and along the coast, as well as giving at least ten talks to varied audiences. It is perhaps not surprising that he did not always respect professionals who claimed to have little time.

The opportunity to write the 1998 Management Plan was perhaps a surprise for Albert Knott who had recently moved to Devon. Because of his experience of chalk ecosystems, he was given the role of Undercliff's Manager. He was less confident about writing plans than about managing the mixture of chalk down and woodland and probably did not fancy the prospect of his involvement after some criticism of the previous plan. Albert did not set ambitious targets, but it was important that his subsequent 2003 – 2008 Plan considered the extent to which his 1998 objectives had been achieved.

Not surprisingly, the natural geological and marine processes had operated unimpeded including damage to South West Water's pumping station following landslipping. Following the designation of the local coast as a Special Area of Conservation, the invertebrates began to feature as important attributes in the way Wallace had considered them 30 years earlier. A study of plant communities at Culverhole was a precursor to similar surveys in other areas of 'soft cliff'. The possibility that communities on Goat Island and the Plateau might benefit from grazing was, again, considered with a feasibility study into the possible use of herbivores having been carried out. Another project, described as limited but useful, looked at the spread of Holm Oak. A new permit system may have helped to safeguard the 'important palaeontological resource'. Also, in the intertidal zone, damage to biological features had been avoided. An important objective, the effective management of areas of chalk grassland, had received more attention and achieved useful results after years of relative neglect. The coast path had been maintained even if the claim that glades *"beneficial to the habitat mosaic had been cut along the path"* was optimistic. The existing sense of wilderness continued, legal obligations had been fulfilled and administrative operations carried out.

Some might say that there had been little actual management and not much planning, but a Manager based a good distance away and left to operate on his own after the loss of two conservation officers, had little chance of achieving much. Despite these difficulties, Albert had made effective contact with the local Conservation Society, had employed botanists to investigate important sites and been a visible influence throughout his first years.

On one occasion, leading a work party to Humble Pond, he left a trail of old socks hanging from branches so that he could guarantee his route on future occasions. On one of these, when heading for the pond with volunteers, a displaying Peregrine enthralled a visitor from Denmark who was unfamiliar with the impressive predator. In my previous book about the Undercliff there is a picture of Albert examining some minute plant on the Plateau while bryophyte expert Roy Jefferey holds a moss to a lens as other enthusiasts look on. Other pictures inevitably show him tending bonfires.

Despite clearance and bonfires, tree cover continued to increase, and Norman noted the disappearance of 20 plant species as light decreased. He also noted fluctuating rabbit populations and their sudden disappearance after doing extensive damage to cliff side fields. With dead Buzzards and Foxes, poisoning was suspected, but tactfully Norman mentioned the possibility of haemorrhagic disease. I had become Chairman of AVDCS, and Norman had passed over to me a thousand sheets of A4 about the Reserve. These became the basis of my 2001 'dossier' completed at Albert's request. Its 700 pages has now become the basis of this rather different account of 250 years of change along this short stretch of Devon coast.

Reports in the *Transactions of the Devonshire Association* on the county's flies, dragonflies, bugs, mosses and lichens are a great source of information but searching for all the local records is demanding and sadly, beyond my patience. I have, however, found some interesting details from visits in the 1990s. In May 1992, a group were disappointed by low orchid numbers but noticed good woodland and calcicole or calcium loving grassland flowers including Nodding Thistle, Ivy Broomrape and Crested Hair-grass. Barbara Benfield identified the tiny liverwort, *Cololejeunea minutissima*, on basic bark and an equally small moss, *Seligera lancifolia*, found among shaded stones. In 1995, there was a reference to Yellow Bird's-nest, an elusive chlorophyll deficient orchid. Tom Wallace had reported it in the early years of the Reserve, another first for Devon, and David Allen later found it near Pinhay in 1968 and, again, in 1996 when he found 12 fruiting plants on a DA visit. Crested Hair-grass was found again, as well as Nottingham Catchfly and Purple Gromwell, its most recent record. In 1998, a 50 year old Walnut on the edge of the Chasm is mentioned.

In addition to the *Transactions*, I have recently found some 'new' Natural England material including a letter from Phil Parr to Albert Knott asking for permission to visit the Undercliff, explaining his reasons, competence and experience of rough ground. I had no idea that he had been coming to Devon, mainly to the coast around Lincombe, for the previous 35 years but of the Undercliff he only knew a section of Charton Bay and the eastern and western coastal ends.

Two moth-men sent their records to Natural England's Conservation Officer, but I suspect their records were actually from outside the Reserve. Roy McCormick, who had caught plenty of moths in our garden, had permission from the Headmaster for five trapping sessions at Allhallows in 1998. His given map reference, 2990, does not give a precise site, and with a need for an energy source for his lamp, I suspect he was outside the Reserve which begins, I think, at the old quarry below the school. Wherever he was, he caught 40 species in May, 86 in June, 60 in July, 113 in August and 24 in September. Not surprisingly, many species were recorded in several months, but he must have had 200 species at least. Ten years earlier S Nash had sent his record of 94 macro and 63 micromoths caught on Ware Cliff to Phil Page.

English Nature's *Invertebrate Site Register* (ISR) for 1995 included two Endangered, two Vulnerable, nine Rare, 23 Nationally Scarce and 19 Local species from the Undercliff. Familiar names among the recorders included Armitage, Ashe, Barns, Bolton, Kennard and Stubbs. The earliest record in the Register was G H Ashe's Adonis Blue in 1924. He had a first for Britain in 1944 when he found a large, 9-13mm, brown weevil, *Hylobius transversovittatus*, on maritime undercliff. Among other esoteric contributions was one for *Strigania maritima*, a red Geophilomorph centipede, normally found on the seashore and, usually according to the site register, in the vicinity of the myriapodologist's home! This one was the first found in fresh water, as reported by Patrick Armitage in number 118 of the *Entomologists Monthly*. Another somewhat specialist record was for the woodlouse, *Cylisticus convexus*, found in 1974 with its map reference (330913) *"on slumping cliffs among loose shale"*, putting it outside the Reserve.

Fox

Buzzard

Barn Owl

Partridge

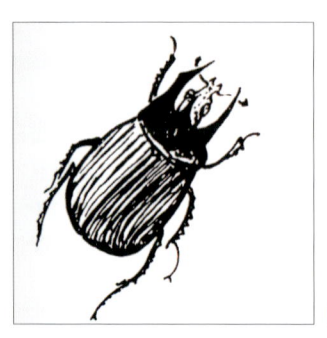

Minotaur Beetle

CHAPTER SEVEN
Birds and Other Animals 1950 – 2003

Once the last Sheep had left the Undercliff, perhaps in the 1930s, Rabbits became the most influential animals there. Before myxomatosis, when grassland was still extensive, they helped to shape the ecosystem. Norman Moore's autobiography *The Bird of Time* referred to his belief that *"the most fundamental changes following the catastrophic decline of the rabbit population were likely to be botanical"*. His colleague, Dr A S Thomas, was to investigate these changes, taking photographs and recording the plants at set points along transects in a range of reserves in 1954 and, again, in 1955. Initially, common downland plants, and rarer ones like many of the orchids, produced a spectacular display but Thomas warned that the more interesting herbs would soon be crowded out by coarse grasses and woody plants. Other changes were inevitable. Moore wanted to study the effect of the disease on six animals including Fox, Buzzard and Minotaur Beetles whose larvae fed on Rabbit dung. This made for an over-ambitious programme but when studying Buzzards, he found that gamekeepers were more important than Rabbits in controlling the predators' population.

After the Undercliff became a National Nature Reserve, it was almost ten years before Hamish Archibald produced a management plan. Before that, and for some time afterwards, there was a non-intervention policy except for some woodland management by Ormsby Allhusen around Pinhay. Contemporary photographs show the extent of surviving grassland but those taken a few years later record the speed of change as scrub developed. When, in 1953, Norman Moore asked F R Smith to describe the local bird populations, he got a detailed reply including a list of summer residents and those which might be breeding. His breeding species included Partridge, Turtle Drove, Cuckoo, Little and Barn Owls, Lesser-spotted Woodpecker, Woodlark, Red-backed Shrike, Cirl Bunting, Tree and Meadow Pipits and Starling.

Exactly what was meant by the 'Landslip area' is not clear but it cannot mean the wide extent of farmland that the species list suggests.

Another observer, J F Cancellor, reported back to the Conservancy two years later with a very different picture. He had only visited the Reserve a couple of times in 1955 but had enough past experience to be able to summarise his knowledge of the area's breeding birds. Wren, Blackbird, Chiffchaff and Chaffinch were 'high density breeders'. He then listed 25 'low density breeders' of which only the Cuckoo would be unlikely today. Also present were Peregrine, Buzzard, Great Spotted Woodpecker, Rock Pipit, Pied Wagtail, Linnet, Yellow-hammer, Bullfinch and two species unlikely to breed there, Swallow and Swift. He expected, but did not record, Long-tailed Tit, Treecreeper and Nightingale.

Part of the difference between the two accounts could be that one had been observing there for years and had, like many birders, accumulated a good list of rarities. However, many on Smith's list were not rare at that time if his 'patch' included farmland. 17 years later when I was counting birds in Wiltshire for the first BTO Atlas, Partridge, Turtle Dove, Cuckoo and Tree Pipit were widespread on farmland and Cirl Buntings were still around. In 1957 Wallace was finding Cuckoo, Lesser Spotted Woodpecker and Starling in the Undercliff with Partridge and Little Owl there in 1961.

Whatever the reason for the different accounts, the 1955 one seems a likely record of the Undercliff birds at that time but neither account sorts out the Nightingale problem! At some stage, Norman Barns maintained that the Undercliff supported a strong Nightingale population, but earlier Devon Bird Reports had no records of the bird there in 1947, 1948 or 1957. A Nightingale census in the county in 1950 *"was not up to expectation through inclement weather"* but 28 observers did manage to locate 63 birds with vague references to Seaton and Axminster. Although Wallace lamented the state of the census work,

The cold winter of 1963 had major effects on bird and fish populations. Snow surrounds voluntary warden Pritchard's house, Lynch Cottage (above) and blocks the road to the Lookout (below).

High Density Breeders

Wren

Blackbird

Chiffchaff

Chaffinch

Periodic Breeders

Nightingale

Lesser Whitethroat

Chapter 7

Peregrine

he did provide a summary of his local experience. Between 1958 and 1969 he had evidence of Nightingales breeding every year except 1963 with pairs at two out of three sites, Charton Bay, Culverhole and the Plateau, in most of the other years. Locally, as over most of Britain, the beginning of 1963 was exceptionally cold even with deep snow in the Undercliff. There, sheltering Snipe and Redwing soon died and over the winter Goldcrest, Stonechat, Nuthatch, Woodlark and Lesser-spotted Woodpecker disappeared; the last two not to be recorded again. What were conditions like in the Nightingales' wintering area at that time?

The Peregrine had been on Cancellor's 'also present' list but by the early sixties that was no longer true. This top predator plays a crucial part in the pesticide story that had been emerging for years. In 1949, Derek Ratcliffe, Peregrine, Raven and, later, mountain plant enthusiast, found broken eggs in an eyrie in Galloway and four years later found four eyries in which at least one egg was broken. By 1961, Pere-grines were declining in the south, a decline which was being linked to the organo-chlorine insecticides, aldrin, dieldrin and heptachlor, which passed up the food chain to accumulate in the brains and livers of the predators. The number of Peregrine territories had fallen from 114 in the thirties to 27. In 1960, concerned about the potential harm to wildlife from these new synthetic pesticides, Norman Moore left his regional work in the South West to take charge of pesticide research at Monkswood Experimental Station. When Ratcliffe sent two addled eggs from Perthshire to Moore, small amounts of dieldrin, DDE (the tissue breakdown product of DDT), heptachlor and lindane were found. This was two years before Rachel Carson's *Silent Spring* which, whatever its failings, alerted the world to the pesticide/wildlife problem. These two years gave Britain a head

Raven

start in confronting the critical situation. Peregrines were not seen again in the Undercliff until 1967.

Back in War time, when some Peregrines were being controlled as predators of the vital message-carrying pigeons, one of the first contributors to Wallace's later lists, G H Ashe of Colyton, found *Hylobius transversovillatus*, a weevil that forms galls on the roots of Purple Loosestrife. These records from 1944 and 1948 were the first for the weevil in Britain. Another indication of the assiduous searching needed by beetle enthusiasts was the finding of *Lycoperdina barstae* which only occurs in puff balls. As Wallace observed *"Collecting is a long and painstaking task involving the sweeping of vegetation, turning over of stones, bark removal, turf cutting etc. Setting the specimens and identification are even more exacting and can be done only by a specialist!"*.

A specialist, but with wide specialisms, was Malcolm Spooner who was to supply Wallace with most of his records of ants, bees and wasps. When he came to the Undercliff in 1964 he found a number of scarce Diptera, two winged flies, including *Frauendfeldia trilineata*, a first record for Devon and a small black wasp, *Diastontus minutus*, which stores aphids in its tiny burrows. He had come primarily to search for sun-loving bees and wasps, characteristic of the soft eroding cliffs, and he found 14 species of these Aculeate (with a sting) Hymenoptera (membranous winged) that day.

The Nature Conservancy's Peter Merrett was an expert on spiders who trapped 65 species in 1965. They were caught in pitfall traps he set up at two sites with 24 at both Charton Bay and Haven Cliff. At the first he caught 57 spiders, 24 harvestmen (Opilionids) and, inadvertently, 12 ants while at Haven Cliff it was 74 spiders, 31 harvestmen and 13 ants. Of the five spider species that he found four were southern, but the fifth, *Widera capitata*, usually occurs at high altitude in the north. *Oxyltida nigrita* and *Lycosa agrestis* are mainly found on chalk as is *Phaeocedus braccatus* which favours stony hillsides. Another species of *Oxyltida, O. blackwalli,* was not known north of Warwickshire. Toddy Cooper updated the spider nomenclature for me twenty years ago but the names may well have changed again since then.

Some of the difficulties in becoming expert in invertebrate identification are suggested in an unlikely essay by John Fowles in his book *Wormholes*. In *The Nature of Nature* he describes enduring jury duty at the Old Bailey in the early 1960s where he heard about a *"diabolically, obscenely nasty"* incest case. Afterwards he went straight to the nearest large bookshop to buy something remote from *"my own insufferably disgusting species"*. He bought Locket and Millidge's British Spiders and then *"for many, too many, years I spent an absurd amount of time peering down an entomological microscope and looking for infinitesimal trichobothria, or trying to decipher, like some insane papyrologist, the exact shape and outline of both male and female sex organs (by which alone the vast majority of smaller spiders can be surely identified). It was during my pursuit of spiders that I was infected by an eventually overwhelming doubt"*. This doubt was about the extent to which he was mesmerised by the binomial system of nomenclature and also about his problem with science in that it tries to dismiss and discount feeling.

Like Merrett the previous year, Dr Speight, with a party of University College London's (UCL's) pioneering conservation course, used pitfall traps; his target was ants, but he also caught spiders in the five inch deep traps which contained an inch of preservative (10% formalin). His traps caught six species of omnivorous *Myrmicine* ants which eat seeds and the secretions of plant bugs as well as being predatory. He also trapped six species of the *Formicinae* which are carnivores of some wide ranging species but also of some relatively sedentary ones that feed off things like root aphids.

The activity of ants is very visible in the Undercliff with active and disused hills all over the grasslands. The species responsible is *Lasius flavus*, the Yellow Meadow Ant, which creates its own microclimate within each nest as described in Davis et al (1992). *"Mounds are built little by little by ants bringing up soil particles in their mandibles from a metre or so deep. These particles are glued to the existing surface of the mound and to plants which continue to grow and thus provide a framework. The coarser sand grains (more than 0.5mm) are less easily transported and so the mound comes to have a disproportionately large amount of fine soil particles compared with the surrounding soil – 84% against 16% in the case of one study"*. Ant hills have a high potassium content, tend to be alkaline and have a low level of organic material. They drain well and only deep rooted perennial plants, like Wild Thyme and Rock-rose, can survive. Any annuals germinate well in the bare soil but most flower quickly before the mound dries out.

Another aspect of ants' nests is that other animals, like the woodlouse, *Platyarthrus hoffmannseggi*, frequently use them. The species feeds on ant faeces and thereby cleans the nest. I have known this small, white, blind woodlouse since 1956 when I was involved in a school project looking at the homing activities of woodlice. The species was not among the seven common ones from four families found by the UCL group. Their most frequent species was *Armadillidium vulgare* which can roll up to conserve water or in self-defence. When it does so, it slots its antennae into grooves in its head. Stephen Sutton, author of *Woodlice* (1974) describes how *"the Pill Bug when attacked by a shrew, seemingly warned by vibration, snaps shut so that the attacker is unable to find a purchase with its jaws and is reduced to pushing the pill bug around by its nose"*.

The numbers and locations of woodlouse families found by the UCL party are shown below in the order of their tolerance of dehydration.

Platyarthrus hoffmannseggi

David Bolton from Exeter Museum found a similar selection in October 1990 with the additions of *Platyarthrus* and a species of *Haplothalamus*, a genus with no West Country records at the time of Sutton's book. The book includes a diagram, here modified by Val Baker (p. 128), showing the trophic relationships of woodlice. It also includes details of an enthusiast who dissected more than 17,000 *Porcellio scaber* to find the parasitic larvae of six species of fly. He found 1,653 larvae of *Parofeburia macula* as a result.

Somewhat larger parasites were among the 43 species of bee on Wallace's species list with the details provided by Malcolm Spooner. Six species in the genus *Nomada* are 'Cuckoos' using the burrows of

Table 6

Woodlouse families in different habitats

	Field Layer	Ground Layer	Bare Ground
Armadillidae	52	162	25
Porcellionidae	70	90	23
Oniscidae (including *Philoscia*)	45	79	10
Trichoniscidae	7	4	0

Chapter 7

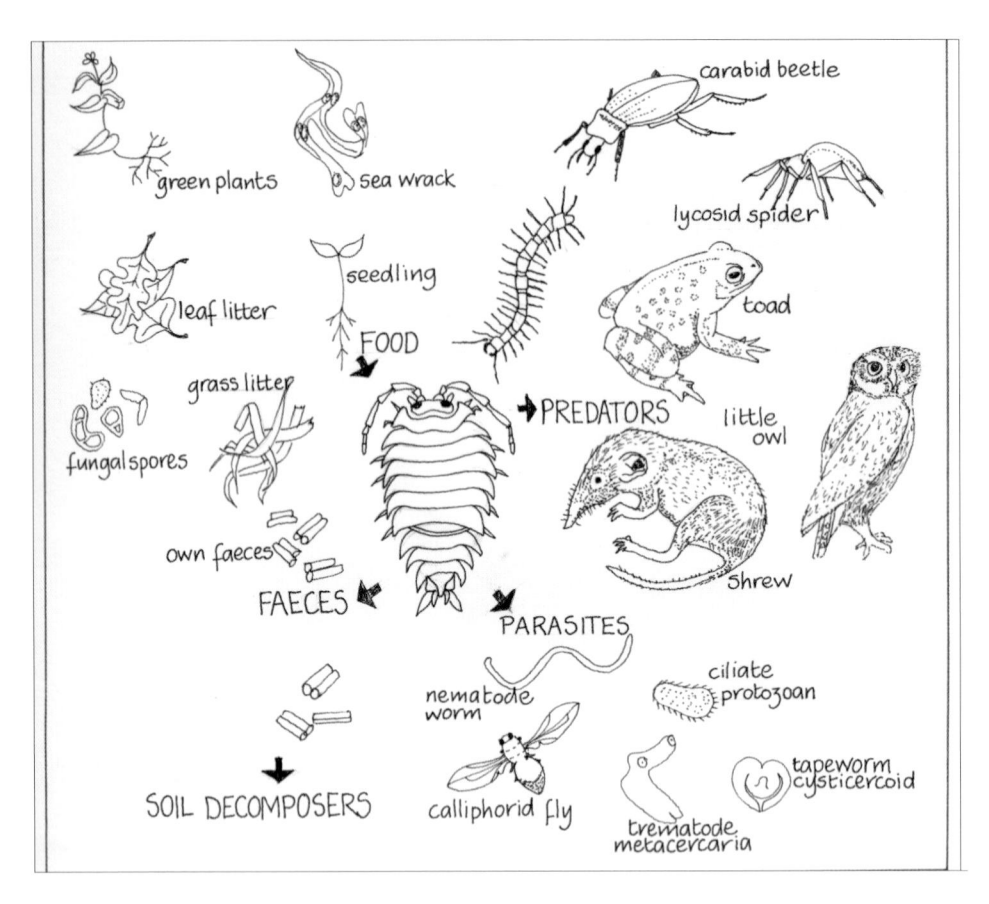

Trophic relationships of Woodlice (Val Baker from Hilary Burn in Steve Sutton's book *Woodlice*)

different sorts of mining bee. Seven *Bombus* bumblebees are social species living in small colonies with a single queen, numerous workers and a few males. They stock their nests with pollen and nectar working longer hours and in worse weather than domesticated bees. The nests are often in the disused burrow of a small rodent. Only young, fertilised queens survive the winter, but they may then be killed by invading parasitic bees of the genus *Psithyrus* which take over the nest and make the bumblebee workers into 'slaves'. These parasites are more powerful than their hosts, have no workers, and females have no pollen baskets on their back legs. Four of these were on the list, as was the bee *Halictus* (now *Lasioglossus*) *angusticeps* which is restricted to the coast between Sidmouth and Purbeck.

Birds and Other Animals 1950 – 2003

Spooner knew Devon well after 43 years working with the Marine Biological Association in Plymouth. He had been a founder member of the Devon Wildlife Trust and was a Fellow of the Royal Entomological Society. In 1979 he was President of the Devonshire Association and as a leading authority on Hymenoptera, he not surprisingly chose to talk about them and about interesting habitats that are often overlooked, in his Presidential address.

"I want to refer to cliff exposures of the softer strata with a sunny aspect. These are much colonised by sun loving insects especially various kinds of bees and wasps which make burrows in the bare slopes or vertical faces, some species preferring sand and some clay. All these Hymenopterous insects have more or less complex behaviour patterns associated with their mating activities. We find both solitary bees and solitary wasps excavating cells from the sides

1. Buff-tailed bumblebee
Bombus terrestris
(10-16mm)

2. White-tailed bumblebee
Bombus lucorum
(10-16mm)

3. Common carder bee
Bombus pascuorum
(10-15mm)

4. Early bumblebee
Bombus pratorum
(9-14mm)

5. Garden bumblebee
Bombus hortorum
(11-16mm)

6. Red-tailed bumblebee
Bombus lapidarius
(12-16mm)

Bombus bumblebees live in small colonies.

and ends of their main burrows and in these they store food for the larvae they will never see. The bees store balls of pollen and honeydew mixture while the wasps hunt some kind of insect or spider prey which they sting to paralyse and transport to a previously constructed burrow... All of these insects, not forgetting their special fly parasites, require well drained ground and as much sunshine as they can get. So, the microclimates provided by coastal sites are important. There is a premium on sunny, sheltered aspects. Unfortunately, these are so often the very slopes that are sought out for human recreation or development of one kind or another".

The animals found in very different, but equally specialised, habitats were described four years later when Patrick Armitage wrote a paper in the Proceedings of the Dorset Natural History and Archaeological Society about the diverse life found in the often transient pools and small streams of the Undercliff. While still at school, he was already excited by water and whirligig beetles. Much later, when he was working with the Freshwater Biological Association, he returned to the Undercliff recording 254 aquatic taxa of which 200 were identified down to species level. His 1983 paper The Invertebrates of some Freshwater Habitats on the Axmouth-Lyme Regis National Nature Reserve was based on collections made in the two previous years but included some records from earlier times. The species were collected from 15 scattered sites, many of which were only transient as landslipping changed the drainage patterns.

He found 115 species of two winged flies including the first British record of *Bryophaenocladius musicola*. Next came 39 beetle species, 22 Caddis Flies and 21 crustaceans including the familiar water-louse and freshwater shrimp, as well as Copepods and Cladocerans *"not intensively studied in this survey"*. There were 13 Oligochaete worms, 12 Hemipteran Bugs and 5 species of flat Platyhelminth worms. Minute Hydracarine mites were well represented and were to form the basis of a later study. There were only common species among his four stoneflies, four mayflies and two leeches and, like others recording in the Undercliff, he did not do well with dragonflies, 5 species, and found only common molluscs. Terrestrial slugs and snails tend to receive even less attention and in the 2013 Bioblitz only two species were listed despite the relative abundance of the two polytypic species of *Cepaea*. Their different variants are favoured under different conditions of natural selection; made famous as a text book example of evolution in action.

The Undercliff has had a few enthusiasts for land-based slugs and snails and, as usual, Norman Barns was among them, recording 34 species. He found glass snails, such as the carnivorous *Vitrina pellucida*, and door snails numerous at woodland edges. One door snail, *Marpessa laminata*, spends its day among the roots of Ash and the night browsing on mosses and liverworts on the Ash trunks. In 1953, Norman Moore commented on the frequency of *Pomatius* and Tom Wallace was attracted by their empty little shells on cliff slopes. Moore found the White-lipped Snail, *Cepaea hortensis*, already mentioned, very common on all grassy areas, and also referred to more local species like the Lapidary Snail found among scrub on Haven Cliff. Most of Wallace's records were due to the UCL party.

Kerney and Cameron (1979) commented that very few species of slug or snail had genuine English names while some of the names that are used are applied to more than one species. Rather than inventing new names, they decided to use the scientific ones for all species. They also observed that the colourless, eyeless, subterranean *Ceciliodes* which favoured ant hills near the cliff edge is most commonly found dead in flood rubbish! The Great-grey and Tree Slugs were recorded by the British Conchological Society in 1969, along with four other species found by the track down to the beach below Rousdon.

David Bolton is the final name along slug recorders. His list included a rather smart species with a yellowish sole and bright yellow to orange marks among 24 species marked off on recording sheets in October 1990. As these only use abbreviations of the scientific names, I found difficulties in working out which species were found that day, but I feel that I am not alone in my ignorance for, on the day of the much later Bioblitz, only a lonely couple of terrestrial gastropods were among the 1,125 listed species. In Table 7 are the names of the land molluscs found by Wallace, Bolton, Barns and others.

Charles Darwin once maintained that the Creator must have been inordinately fond of beetles because of their diversity and number. Tom Wallace, however, only listed 90 species. In 1989, that number was challenged by the 74 found by Adrian Turner whose identification difficulties were exemplified when his hoped for Red Data Book species turned out to be more common. Wallace's suggestion that there might be 900 Undercliff beetles got some support from the fact that only six were on both his and Turner's lists; two ground beetles, the rare *Stenus*

Chapter 7

Table 7 - Land Molluscs listed in the order used by Kerney and Cameron

Abida secale – Large Chrysalis Snail	W	*O. helveticus* – Glossy Glass snail	W
Pupilla muscorum	B	*Limax maximus* – Great Grey Slug	C
Lauria cylindracea	B	*L. marginalis* – Tree Slug	C
Vallonia costata – Ribbed Grass Snail		*Euconulus fulvus* – Tawny Glass snail	C
A tiny species found at Dowlands	W	*Cecilioides acicula*	B
Acanthella aculeata – Prickly Snail	WB	*Cochlodina (Marpessa) laminata*	
Ena obscura	B	Plateau Door snail	W
Discus rotundatus – Rounded Snail	W	*Macrogastra rolphii* – Rolph's Door Snail	C
Arion ater – Large Black Slug	C	*Clausilia bidentata* (rugosa)	B
A. Subfuscus	B	*Candidula intersecta (Helicella caperata)* Wrinkled Snail	WB
A. fasciatus Bourguignat's Slug	C	*Cernuella (Helicella) virgata* – Banded Snail	WB
A. hortensis Garden Slug	C	*Helicella itala*	B
A. intermedius Hedgehog Slug	C	*Cochlicella acuta* – Pointed Snail	P
Vitrea crystallina – Crystal Snail	P	*Trichia hispida* – Bristly Snail	W
Aegopinella (Retinella) pura Clear Glass snail	W	*T. Striolata* – Hairy/Bristly Snail	WB
		Heligonia lapicida - Lapidary Snail	B
A. (R) nitidula – Smooth Glass snail	W	*Cepaea nemoralis* – Brown-lipped Snail	W
Oxychilus cellarius	B	*C. hortensis* – White-lipped Snail	WB
O. alliarius – Garlic Glass snail	W	*Helix aspersa* – Common Snail	WB

Key: B = Bolton   C = Conchological Society   P = Mrs Peters   W = Wallace

Additional species include the very common *Pomatius elegans* recorded by both Wallace and Bolton, *Carychium minimum* and *cochlicofa lubrica* (both Wallace) as well as *C. lubricella* (Bolton, who does not attempt English names).

guttula, from wet places, two Lucanids and the 24 Spot Ladybird. The Lucanids are relatives of the Stag Beetle, the Small Rhinoceros Beetle and the Lesser Stag Beetle.

Table 8– Turner's Millipedes (1990)

Cylindroiulus punctatus	Haven and Bindon Cliffs and Dowlands
C. britannicus	A Snake Millipede found at Dowlands
Blaniulus guttulatus	A Spotted Snake Millipede from Bindon
Ophyiulus pilosus	Bindon and Dowlands
Tachypodoiulus niger *Proteroiulus fuscus*	Snake Millipedes from Bindon

Striped millipede

The UCL party had only recorded one millipede, *Glomeris marginata*, which like the woodlouse mentioned earlier can roll up to protect itself. Turner collected six species which were identified by the British Myriapod Group who also confirmed his nine centipedes including five species of *Lithobius*.

If you look at the generic names in Table 8, you may understand why, in 1658, it was deemed that *"unless they have many feet they cannot be numbered or named among the Juli"*.

In 1993, I was to do a lot of numbering and naming when I started to survey the birds in parts of the Undercliff. As soon as we had moved to Devon I contacted John Woodland, then the British Trust for Ornithology representative for the county, and he immediately suggested some sort of census work in the Undercliff and a meeting with Norman Barns. One February morning I met Norman and he took me through the Chasm and over Goat Island, and when I appeared fascinated by the area he then took me to the Plateau. On our way there a spectacular cliff fall clattered behind us and scattered the deer but exciting though that was, I felt that Goat Island and the Chasm would make the ideal area for a Common Bird Census (CBC).

Although Haven Cliff was not accessible at that time, birds at the eastern end, within the census area, could be seen from the coast path above as the trees and shrubs which now obscure the view were very much smaller.

Altogether 38 species bred, as, in addition to those in the table Buzzards had nests every year and Stock Dove, Tawny Owl, Jay, Whitethroat and Lesser Whitethroat bred in two of the five. Jackdaw breeding on the inland cliffs were outside the census area.

In winter the total was lower with the 19 in the table plus Buzzard, Peregrine, Great Spotted Woodpecker, Raven, Magpie, Jay, Stonechat, Grey Wagtail, Redwing and Greenfinch.

In 2021 a new venture was to attempt some ringing in the Reserve. Initially four possible sites in the Chasm were investigated but later Naomi Brookes chose a more open site on woodland edge above the Chasm. Early catches included plenty of Chiffchaffs one of which had been ringed in Spain, as well as other migrants including Redstart and Grasshopper Warbler. Later in the year, in October, about half the twenty-odd birds she caught were Blackcaps. They were all quite heavy so had no doubt been feeding well before heading west ready for the short crossing to France. Nearby, but not in the Undercliff, an Ortolan Bunting had attracted enthusiastic birders a few weeks earlier.

In 1995, in contrast to Wallace and his associates, Mike Edwards had only five days of far from perfect weather to search for, and find, a number of invertebrates new to the Reserve. He concentrated on aculeates and selected groups of flies (hoverflies and the larger *Brachycera*) with casual records of Orthoptera. Despite July being too late for many species, 156 were recorded. One strip on Haven Cliff, about 300 metres long, was good enough to be visited twice. One coastal seepage provided three out of the four records of Soldier Flies including the nationally scarce *Stratiomys potamida* and *Oxycera pygmaea*. Among other records made, there was the earwig *Forficula lesnei*, new for Devon, and four Red Data Book aculeates including large nesting aggregations of *Lassioglossum laticeps*, in slumped mud at the base of the cliff.

Woodland was only surveyed when the recorder was moving from one site to another but the hoverflies *Fernandinea cuprea* and *Sphegina*

Birds and Other Animals 1950 – 2003

Table 9a
Breeding Birds Totals in the five years (1994 to 1998) of the Common Bird Census (omitting coastal species).

Wren	92
Chiffchaff	81
Blackcap	80
Robin	69
Blackbird	60
Chaffinch	36
Great Tit	33
Wood Pigeon	32
Blue Tit	31
Song Thrush	30
Goldcrest	30
Bullfinch	28
Dunnock	23
Linnet	20
Marsh Tit	16
Yellowhammer	15
Pheasant	11
Longtailed Tit	10
Coal Tit	10
Spotted Flycatcher	8
Nuthatch	8
Tree Creeper	7
Carrion Crow	5

Table 9b
Number of birds counted in the five years (1994 to 1998) on four winter visits (omitting coastal species).

Wren	55
Blackbird	55
Robin	33
Bullfinch	31
Longtailed Tit	27
Blue Tit	26
Song Thrush	24
Coal Tit	23
Great Tit	22
Chaffinch	18
Goldcrest	15
Jackdaw	15
Wood Pigeon	14
Pheasant	13
Dunnock	12
Nuthatch	11
Carrion Crow	11
Marsh Tit	10
Tree Creeper	6

Blackcap (L), Robin (R)

Chapter 7

Clockwise from top left:
Song Thrush, Great Tit, Jay,
Treecreeper, Bullfinch

clunipes, one widespread but infrequent, the other widespread but uncommon were found despite the relative youth of the woodlands. It was felt that they would benefit from additional open glades with standing dead timber as many species exploit the interface between woodland and more open spaces at some stage in their lives.

RDB species are on the Red Data Book lists of endangered species with book one, species found only on coastal soft cliffs, book two, those associated with soft cliff and book three, species found in a wider range of habitats. Nationally scarce species are only found in a small number of 10km squares.

In 2002, invertebrate recording got a boost with the establishment of the Invertebrate Conservation Trust, widely known as 'Buglife'. Five years later the book *Managing Soft Cliffs for Invertebrates* was published. Its author, Andrew Whitehouse, was able to include the findings of a 2003 survey carried out by David Gibbs who had recorded 292 species of which 22% were either on the Red Data Book list or nationally scarce. His results, together with the earlier ones of Edwards, were summarised in the Buglife book. It had been found that the coast between Haven Cliff and West Bay supported more soft cliff species than any comparable length of coastline in Britain. Of the 29 invertebrates found only on soft cliff sites, 18 were found on this stretch of coast. In addition to these, Grade 1 UK Biodiversity Action Plan (BAP) species, there were 9 Grade 2 and 23 Grade 3 species. Grade 2 insects are strongly associated with soft cliffs while those in Grade 3 also occur in other habitats.

Both the east and west coasts of the Isle of Wight, with over 30 species, and the equally favoured south Devon and east Dorset coasts were among the most productive soft cliffs. In Wales, despite much hard rock, the north and south coasts of the Lleyn Peninsula have most species, while Gower and Ceredigion also do well, as do Yorkshire and the Castlemartin Peninsula in North Devon. Of Grade 1 species, limited to soft cliffs, eleven were Coleoptera, six each of Lepidoptera and Hymenoptera and five Diptera with a single Hemipteran, the Shore bug, *Saldula*. The Buglife book maintains that these cliff habitats have been undervalued and that nature conservation effort in the UK has been driven by interest in plants and vertebrates. Bare ground is of key importance to invertebrates on soft cliffs and other habitats. Bare areas are favoured hunting grounds for visual predators. Specialist 'pit predators', such as the larvae of the Cliff Tiger Beetle also favour bare ground where they wait in burrows to ambush prey.

Table 10 – The Hymenopteran Records of Wallace and Edwards

Taxonomic Groups	Number of Species	
Wasp Families	Wallace	Edwards
Chrysididae (Brilliantly coloured cuckoo wasps)	4	2
Methocidae (Including nationally scarce *Methoca ichneumenoidea*)	2	1
Pompilidae (spider Hunters including RDB 2 *Cryptochelius notatus*)	4	5
Eumenidae (Mason wasps making cells in holes in the ground, wood and masonry)	4	3
Vespidae (Social Wasps including the Hornet)	2	4
Specoididae (Including two nationally scarce solitary wasps)	12	21
Solitary Bee Taxa		
Colletidae (Characterised by wasp-like mouth parts)	6	5
Megactilidae (Leaf cutter bees placing eggs in rolled up leaves)	0	3
Hallictidae (Including RDB2 *Lasioglossus augusticeps* and *L. laticeps*)	3	5
Nomada (Including *N. fucata*, nationally scarce and *N. fuloxornis* RDB 3)	4	0
Psithyrus (Like *Nomada*, a genus of cuckoo bees)	10	9
Andrena (Mining Bees including *A. hathorfiana*, RDB3, *A. riparia*, nationally scarce and *A. simullina* RDB2)	8	9
Social Bees		
Of the genus Bombus or Apis	8	6
Ants (Including *Leptotorax tuberum*, nationally scarce A)		
Omnivorous Myrmicine species	6	3
Predatory Formicine species	6	4
Total:	79	80

"*Friable bare ground*" continues the Buglife report *"offers nesting sites for burrowing bees and wasps. Solitary bees provision their nests with pollen and nectar while the wasps collect prey items (such as weevils) – depending on the species. The most suitable substrates are sufficiently friable to allow burrowing, but firm enough to prevent burrows collapsing. Films of water running over mud and mosses provide important breeding sites for flies and beetles"*.

125 years before Buglife, and almost 100 before Wallace and his lists, F O Morris's *History of British Butterflies* referred to Silver-studded Blue and Brown Argus on Pinhay Cliffs which would have been very different in the 1870s. It is the only reference to the blue in or near to the Undercliff, but the Brown Argus has a well documented history. This includes a scatter of five 1980 records in *Devon Butterflies* (1993) but the Butterfly Conservation database also has local records since 1997. Wallace's 1967 booklet on Undercliff wildlife describes the Brown Argus as *"frequent among the Rock-roses in 1933 but as having become less common"*.

Table 11 – Orthoptera on Wallace's List and/or recorded by Edwards

Comments are mainly from Wallace

Number of Species		
Bush Crickets or Long horned Grasshoppers (*Tettigoniidae*)	**Wallace**	**Edwards**
Concephalus discolor (Long-winged Conehead). A species increasingly widespread in Southern England.		E
Leptophyes punctatissima (Speckled Bush-cricket). Described by Wallace as a strange, fragile spider-like creature.	W	E
Meconema thalassium (Oak Bush-cricket). A nocturnal tree dweller sometimes swept up in woodland.	W	
Pholidoptera griseoptera (Dark Bush-cricket). A brisk chirper by day and night. Hind wings absent.	W	E
Tettigonia viridissima (Great Green Bush-cricket). Bright green. It may stridulate loudly by day or by night.	W	E
Grasshoppers or Short-horned Grasshoppers (*Acrididae*)		
Chorthippus albomarginatus (Lesser Marsh Grasshopper). A southern species locally common in west grasslands.		E
C. brunneus (Field Grasshopper). A good chirper who flies well and is very common.	W	E
C. parallelus (Meadow Grasshopper). A non-flier and modest songster. The commonest species.	W	E
Ground-hoppers (*Tetrigidae*) with much reduced wings hidden by a hood		
Tetrix undulata (Common Ground-hopper). Grassy, shrubby places, e.g. Pinhay Cliffs.	W	
T. subulata (Slender Ground-hopper). Very plentiful 17/04/55 (W. Hooper). Many in dense cover, Culverhole 03/10/64.	W	
T. ceperoi (Cepero's Ground-hopper). On bare ground April 1966 (M Speight), Haven Cliff 26/10/90 (D Bolton).	W	

Chapter 7

Table 12 – Nationally Scarce and RDB Species Found in the Undercliff
Edwards (1995) and/or Gibbs (2003)

Species	Status
Cliff Tiger-beetle (*Cylindera germanica*)	RDB 3 / UK BAP Grade 1
Shorebug (*Saldula arenicola*)	Na / Grade 1
Mining bee (*Andrena simellima*)	RDB 2 / Grade 2
Mining bee (*A. spectabilis*)	Na / Grade 2
Mining bee (*Lassioglossum laticeps*)	RDB 2 / Grade 1
Mining bee (*L. angusticeps*)	RDB 3 / UK BAP Grade 1
Nomad Bee (*Nomada fucata*)	Na / Grade 3
Cranefly (*Dicranomyia goritiensis*)	RDB 3 / Grade 2
Cranefly (*Helius hispanicus*)	RDB 1 / Grade 1
Micromoth (*Sorobipalpula tussilagiris*)	RDB 1 / Grade 1
Chloropid Fly (*Platycephala umbraculata*)	RDB1 / Grade 1

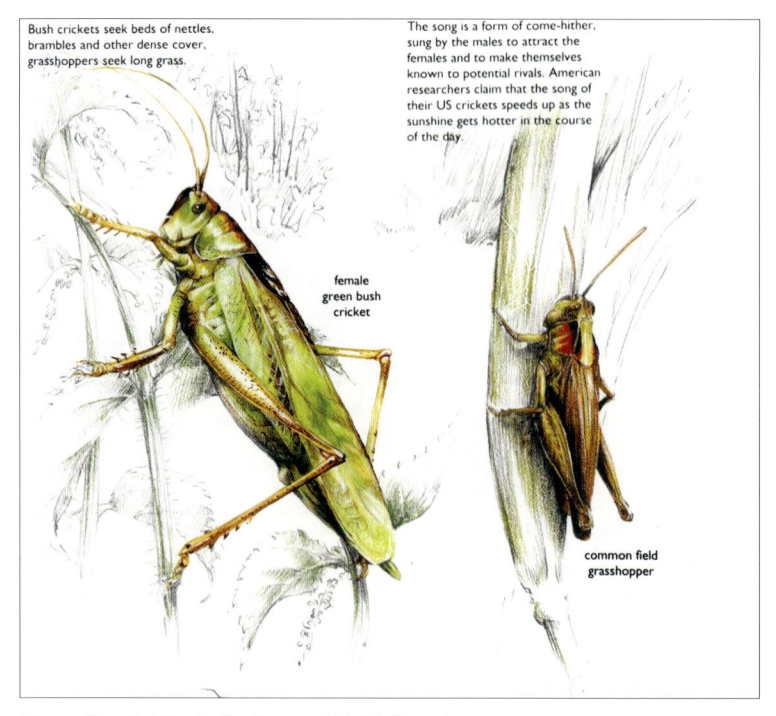

Elaine Franks' Bush Cricket and Field Grasshopper are from Geoffrey Young's *Watching Wildlife*

Cliff Tiger Beetle

With the great changes in both habitats and butterfly populations since Morris, and considerable ones since Wallace, the fritillaries are a group which has suffered. Post-1980 records of Silver-washed are in all appropriate tetrads in the 1993 book but there are no High-browns since 1930, no Small Pearl-bordered since 1956 or Dark-green since 1966. I photographed a rather battered Dark-green in 2001 and there have been enough recent records to suggest that a small remnant population may survive in or near the Undercliff. Norman Barns found the Pearl-bordered quite frequently until the number of Violets fell as tree cover, particularly of dense Holm Oak, increased.

The database has a selection of Wood White, Dingy Skipper and Green Hairstreak records from the 1990s but not a single Grayling, Grizzled Skipper or Small Heath, whereas in 1967, the first of these was *"fairly common each year in open grassy places"*, the second *"quite common in open grassy places"* and the third also common in *"grassy places"*. Twenty of another scarce species, the Chalk-hill Blue, were seen by Wallace a few feet outside the Reserve, on the clifftop at Whitlands, in September 1978, a site from which they have been reported both before and since and butterfly man, Phil Parr, believes that there may be a small colony on the steep chalk cliff face below the Plateau where Kidney Vetch grows.

Mining bees, wasps, beetles and a Glanville Fritillary among wet slumps dominated by horsetails and Birds-foot Trefoil. (Buglife).

CHAPTER EIGHT
Soft Cliffs and Managed Grasslands

The 1998-2003 Management Plan states that *"the more mobile soft cliffs show a complex sequence of successional communities related to degrees of instability and the angle of slope. The vegetation of these sites forms a mosaic of pioneer, ruderal, grassland, scrub and woodland communities. Streams and flushes provide a freshwater element and seepage lines may be rich in orchids"*.

This is true of Pinhay Warren where, among the pools near the cliff edge, are desolate patches of damp, grey clay with sandstone blocks and islands of vegetation. These move down the slope during and after winter rains. Many of them are covered with Narrow-leaved Everlasting Pea and the inevitable colonists Hemp-Agrimony and Buddleia with their wind spread seeds. Blue Fleabane, also wind dispersed, is all over the place with isolated plants of Centaury and Yellow-wort.

The Warren provides one extreme of the wide ecological variation to be found along the coast which, from Sidmouth to West Bay, has been designated as an SAC or Special Area of Conservation. The instability of many of the cliffs provides ideal conditions for invertebrates but presents challenges for plants. Because of the opportunities for colonising the bare, exposed areas of varied soils, a range of ant, bee, wasp and beetle species flourish and provide the reason for the designation. The Warren, together with Culverhole and the cliffs above Charton Bay, is one of the best examples of the soft cliff habitats which were investigated by Michael Cooke and David Gibbs of South-West Ecological Surveys in 2002 and 2003.

The extensive subsidence at Pinhay only began relatively recently; it had previously been a popular picnic site in (early) Victorian times. Much of the area is covered in sandy or dark mudstone silts which can easily be moved downslope where very wet conditions, with flowing

Chapter 8

water, are frequent. The loosened material eventually falls over the cliff edge. This erosion is constantly promoted by tide and wave action at the cliff base, perhaps helped by old quarrying activities. No doubt the development of tree cover at the top of the Warren has not only changed the drainage patterns but also affected the weight bearing capacity of the shallow soils and led to the extensive movements seen in the last sixty years.

Near the cliff edge, mosses and locally abundant liverworts characterise the very wet silty substrate where Yorkshire Fog and Hoary Willow-herb form part of the pioneering community. Higher on the Warren, Sycamore, with a dense Bramble understory, dominates but many of the trees are sliding down, towards the sea, breaking up the woodland canopy and creating opportunities for colonisation by the Everlasting Pea, Marjoram, Wild Strawberry and Wood Sage. Dead and dying Sycamore surrounds areas of Harts-tongue Fern, Madder and Tutsan with pioneers including Ploughman's Spikenard and Hoary Ragwort moving in among the low growing scrub. Very little of the vegetation cover, where any exists, fits satisfactorily into the categories of the National Vegetation Communities (NVC). This variation from wider UK vegetation types seems to reflect the unique geology and associated instability of the site, as pointed out by Mike Lock and David Allen in the *New Flora of Devon*.

Painting of the Southern Marsh Orchid. The painting by "S.H." was bought at a Wiltshire Trust exhibition in Salisbury in 1988.

The different areas of pioneering vegetation on *"loose, freely eroding, silty, sandy substrates"* present a common pattern. Parts of Culverhole are as unstable as the Warren and have been so for a longer time. The slopes are steeper and fall directly to the sea rather than the cliff edge. Other parts have been stable for many years and the dense scrub is only penetrated by a narrow, somewhat hazardous, route to what Phil Parr, in his 2006 butterfly report, calls the West Fen. This is separated from the East Fen by more scrub and Willow cover with much slumping and already slumped Grey Willow. Phil believed, rightly, that at least in the short term *"the constant slipping of the past fen grassland towards the shore will be enough to keep some of it clear of scrub"*. Recently, some of the flowers at the base of the western area were Pale St John's-Wort, Southern Marsh Orchid, a few flower-heads of Marsh Fragrant Orchid, Black Bog-rush and a rapidly decreasing population of Bog Pimpernel. With the very dry conditions there at the time, the decrease is not surprising. The much wetter East Fen had some 50 spikes of Marsh Helleborine and a single large flower head of the Fragrant Orchid in 2020.

Back in 1993, W and J Cox had found *"the productive and diverse flushes"* undergoing threat from slumping into the sea and from encroachment of scrub, particularly Gorse. They found 65 vascular plants and 21 bryophyte species. As their survey was in late summer, it is not surprising that on a visit earlier in the year in 2001, David Allen and Roy Jefferies added another 22. It is much more surprising that David, who had botanised at Culverhole for years, had never previously visited the eastern area.

At Charton Bay, Cooke and Gibbs describe three separate areas of open failed slope on the lower cliffs. They are among patches of dense scrub which are periodically on the move and the complex patterns that they described in great detail will certainly have changed. Their polygon 1 was adjacent to the shoreline at the west of the site with polygon 3 immediately upslope supporting, in 2003, established calcareous grassland. Polygon 2 was to the east and *"represented a section of failed and actively eroding slope exposing much of the local geology"*. Clumps of Pampas Grass dominate the comparatively stable belt above the decidedly unstable slopes above the shore. Higher up, polygon 1 is a sparse pioneering grassland dominated by Wood Small-reed with Tall Fescue, Blue Fleabane, Black Mustard, Wood Sage, Smooth Hawksbeard, Madder, Marsh Thistle and Ragwort. Scattered stands of Grey Willow and Holm Oak occur throughout.

The shallow lower slopes of the second polygon were dominated by Wood Small-reed with abundant Great Horsetail and frequent Holm Oak, Privet, Wayfaring Tree and Grey Willow. Above it, a mix of eroding mudstone, silt and terraces of slipped calcareous grassland forms a quarter of the polygon, again, with Great Horsetail and locally frequent Bird's-foot Trefoil. Another 35% on the upper slopes of limestone scree has a varied composition of Mouse-ear Hawkweed and Bird's-foot Trefoil with frequent Tall Fescue. Other grasses and Dwarf Thistle are locally frequent. The slopes at the top have invading scrub of Holm Oak and Buckthorn together with Dogwood, Wayfaring Tree, Blackberry

Chapter 8

and Hawthorn. Once again, the communities have little affinity with any NVC category.

East of Charton Bay, on the other side of Humble Point, is an area which old aerial photographs show as open cliffside but which by 2007 was colonised by Holm Oak. This was removed and the year after the slope had been cleared. David Allen, working with Joan Millard and Marjorie Waters, attempted to record the sequence of events which followed the clearance. They scattered five one metre quadrats at random with a sixth on the remaining woodland as a control. On four occasions in 2008 and, again, on 6 May 2009, the areas of bare ground, litter and vegetation was assessed. At the start in April 2008, 85% of the surface was bare and only 15% vegetation covered, but a year later, 75% was vegetated with a striking change in diversity. David's report in the *Axe Vale Conservation Society Magazine* tells how the plots that had contained one or two species at the start now had up to 16. Many of those that had appeared had wind dispersed seeds including aggressive colonisers like Sow Thistle, Groundsel, Creeping Thistle and species of Willowherbs but the appearance of Yellow-wort, Hairy Violet, Hawkweed Oxtongue, Ploughman's Spikenard, Glaucous Sedge and Bird's-foot Trefoil suggested that the seed of at least some typical chalk grassland plants had persisted in the soil. If the Giant Fescue dominated two of the quadrats and emphasised how much management would be needed over the years, the collapse of the whole area, as the soft cliff subsided into the sea, provided another reminder that in many parts of the Undercliff, the only permanence is perpetual change.

On much of Goat Island and the Plateau, change is less noticeable but the steep slopes around the one, and the sea cliff at the other, are far from stable. At both sites, trees adjacent to the grassland provide opportunities for colonisation. As there was a non-intervention policy

Yellow-Wort

Harebell

for many years after the designation of the National Nature Reserve, the vegetation on the few areas that survived as grassland was very different from what is there now.

That was clearly shown in 1966 when University College, London organised a field studies week as part of their recently established Conservation course. On their visit to the Undercliff, geographers and zoologists outnumbered botanists but even early in the year, March into April, plants were not neglected, and profile transects across two sites were attempted. Both crossed two distinct communities with one sampling the vegetation in Ash-wood, south of the coast path and across the Plateau and the other, somewhat optimistically, crossing the Chasm and Goat Island.

On the Plateau, grassland described as 'tussocky' appeared to be dominated by Salad Burnet which was recorded in all ten of their three foot squared quadrats. Wood Sage was found in eight quadrats, "Thistle" in four and Bramble in three together with some Dogwood, Wayfaring Tree and Ground Ivy. More than 20 years later, in July 1989, still with no management, Clare Freeman, from Royal Holloway College, attempted an ecological evaluation of the Undercliff with the aim of finding a way to prioritise sites for management within a single reserve.

Her project involved a detailed survey of seven sites and was *"perhaps the most detailed study which had been undertaken"*. Her report extended to 87 pages and, like the 49 page appendix, had its complications. Her ranking of the seven chosen sites differed markedly from that of Warden Norman Barns who helped with her research. Norman made the Plateau the top priority for management with the birchwood at Whitlands second, while Clare's top site was the Chasm

Eyebright

Early Gentian

Chapter 8

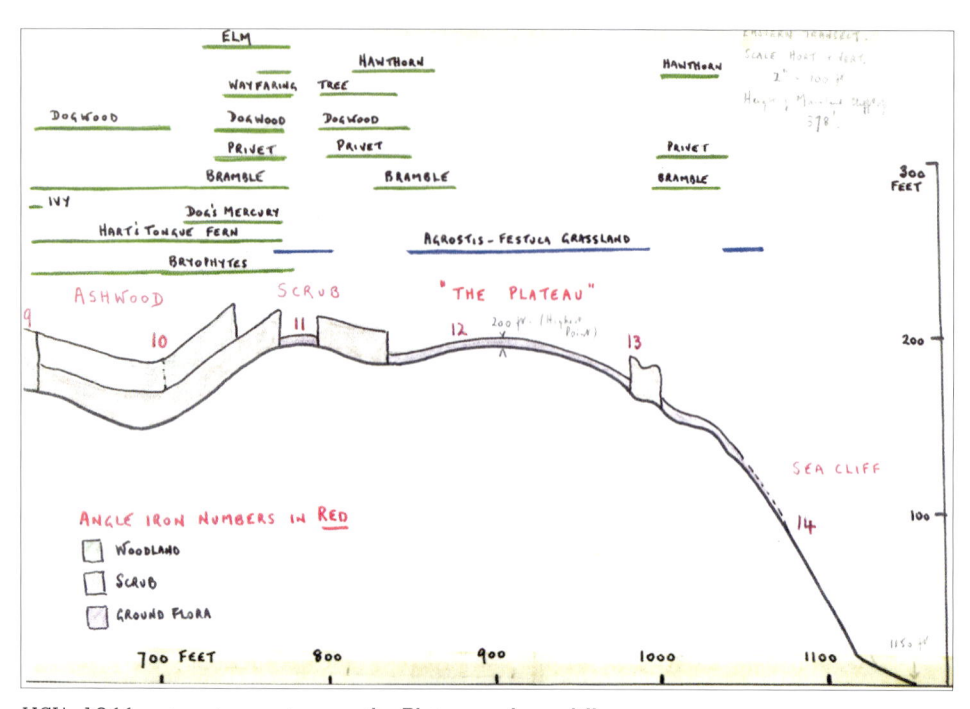

UCL's 1966 eastern transect across the Plateau and sea cliff.

Norman made the Plateau the top site for management.

Clare's top site was the Chasm, photographed earlier by J. Jesty.

followed by the coast path birchwood. Next came Ware Common, The Clearing and nearby Hazel Grove, with the Plateau and Hazel Grove itself, at the bottom. Her ranking was based on plants alone and much of it now seems peculiar but, as we have seen, the Plateau was then unmanaged as part of the non-intervention policy. She found five species in all her Plateau quadrats; Wood Sage, Silverweed, Ladies Bedstraw, Yarrow and Cocksfoot. Wood Spurge, Creeping Cinquefoil, Dogwood, Ribwort Plantain and Bristly Oxtongue were in four and Bramble, Wild Basil, Black Medick and Self-heal in three. Silverweed, Creeping Cinquefoil and Black Medick could well be considered ruderal, a word the dictionary defines as *"growing in or among stone, rubbish or peculiar to rubbish heaps"* so the abundance of the three species strongly suggests the need for management and their disappearance by the time Jonathan Cox investigated the area in 1993 points to the significance of the mowing regime that had been introduced. The mowing must also have done much to reduce Dogwood and Bramble as well as causing the disappearance of the woodland plants, Wood Spurge and Gladdon.

Once mowing and raking-off had been introduced, vegetation change was rapid, as shown by differences over the ten years between 1993 and 2003. Jonathan Cox found only 24 species in 1993 with False Brome dominant but 45 species ten years later when plant height had halved. Bird's-foot Trefoil, Eyebright and Ribwort Plantain had greatly increased, as had False-oat Grass and Creeping bent. Wood Sage had disappeared but Yellow-wort, Centaury, Dwarf thistle, Rock-rose, Ploughman's Spikenard, Self-heal and Red Clover had colonised or increased sufficiently to be recorded.

Since it slid from the cliffs above, hundreds, or even thousands, of years ago, the Plateau has presumably persisted as some sort of grassland, but Goat Island's recent history is more dramatic. Farmed until 1839, when the landslip led to its isolation, growing crops again in the Second World War and then developing increasing scrub before clearance, to varying degrees and in different parts, started conservation management. This is now mainly carried out by a small army of workers active for three days in September. They cut and rake off much of the vegetation taking care to avoid the ant hills which are often individually treated.

Although the field work of Cooke and Gibbs in 2002 and 2003 was to *"carry out detailed survey of areas of active landslip"*, they did study two distinct parts of stable Goat Island but as they did not extend their study to the Plateau, it is convenient to describe the plants of both

Chapter 8

areas in the words of a contribution to the 2016 *Flora of Devon*. There, David J Allen and Mike Lock maintain that *"Species rich chalk grassland in the Undercliffs is best represented on Goat Island and the Plateau where in excess of 30 species per m² are usually found"*. The unusual management of these sites makes for unusual vegetation: although Fescues, the Oat grasses, *Avenula pubescens* and *A. pratenris*, and Yellow Oat-grass (*Tristetum flavescens*) are present, species typical of grazed grassland are absent; the frequent presence of False Brome (*Brachypodium sylvaticum*) and the occurrence of many woody seedlings tends to suggest that the grassland is, in fact, scrub.

Nevertheless, many 'classic' chalk grassland herbs are found... Harebell, which inexplicably is a very rare plant in South West England, occurs only on the Plateau, which was also the stronghold of the rare annual endemic Early Gentian, the abundance of which has varied from year to year. 19 plants were counted in 2006 on shallow soil at the cliff edge where the scarce Soft-brome also occurs.

Earlier counts of the Gentian were 300 in 1998, after extensive disturbance due to scrub clearance, and 30, 28, 57 and 6 in the years up to 2002 with 13 in 2005. In *Britain's Rare Flowers* Peter Marren describes it as *"a peculiarly capricious plant"* and maintains that its seed *"is*

Greater Butterfly Orchid - Autumn Ladies-tresses - Twayblade

vitalised by the right conditions, namely, disturbance and a wet winter followed by a warm wet spring".

The *Devon Flora* describes "*how the Autumn Gentian can be especially abundant on Goat Island where in some seasons there are at least 30,000 plants. Nine orchid taxa are found in the grassland of Goat Island including Bee, Pyramidal, Greater Butterfly and Autumn Lady's-tresses, of which there are sometimes more than a hundred plants. Rather surprisingly, both Marsh Helleborine and Southern Marsh-orchid, with its variety junialis (Leopard Marsh-orchid), also occur here in open, dry chalk grassland.*"

In a 2020 *Conservation Society Newsletter*, David was, again, writing about orchids, describing some of their pollination mechanisms. Twayblade, particularly common on Goat Island East, is adapted for pollination by unspecialised insects, particularly Ichneumon-wasps, but also small flies and beetles. The plant has its nectar well exposed and a visiting insect, feeding on the nectar, gets its head covered in a glue exuded from a structure called a rostellum which sticks the pollinia to the insect's head. The mechanism used by the Pyramidal Orchid is more sophisticated. It has a lip with two conspicuous projecting ridges which leads moths and butterflies to a nectar-containing spur *"like an aeroplane being guided on a runway"*. The Greater Butterfly with its white flowers is adapted for attracting night flying moths. Around the Mediterranean, Bee Orchid relatives are sufficiently bee-like to attract insects, whereupon pheromones induce the males to attempt copulation. The column of the orchid immediately descends and glues the pollinia to the bee's head before it moves on to pollinate the next flower. Despite all this sophistication, British bees hardly ever visit the orchids which still manage to set seed, evidently accepting their own pollen.

Common Spotted Orchid

Orchids were very scarce on Goat Island East in 2001, after Phil Page had opened up a route to the tiny area which had survived in a scrub-free state, Twayblade was the only one recorded by Cooke and Gibbs. In their quadrat 1, in the patch of established grassland, Rough Hawkbit was dominant, Glaucous Sedge abundant, and Tall Fescue, Quaking

Chapter 8

Ivy Bees colonised parts of Glade 2.

Grass, Eyebright and Bird's-foot Trefoil frequent, whereas False Brome and Bracken dominated the neighbouring quadrat which had only recently been cleared of scrub. Tall Fescue and the sedge were, again, abundant, as were Wood Sage and Dog Violet. Evidence that the established grassland had long been there came from the size of the ant hills but even so, its plant community was very different from other, superficially similar, sites on Goat Island. Shallower soil, past land use and its tree-enclosed position could account for this. Whatever the causes of the differences, the changes have continued at both sites, which are now more similar as even the ants are spreading into the newer grassland. Hemp-agrimony and Ploughman's Spikenard now thrive on the margins of both plots. The Bracken, constantly pulled or cut back, is much reduced, the ant hills have produced habitat for Lady's Bedstraw and Marjoram, while orchids are now so common that one must tread carefully. There can be up to a thousand Common-spotted, a hundred Twayblade, a good number of Pyramidal and a variable count of Bee Orchids, mainly on the old grassland. Most grasses and sedges have decreased whereas Carline Thistle, Everlasting Pea, Fleabane, Small Scabious and a few Autumn Gentian have increased the diversity.

It was also in 2001 that bryophyte experts Mark Pool and Roy Jefferies were in the Undercliff for a couple of days to record some of the mosses and liverworts that had not been studied for some 40 years. On the second day out, a navigation error brought us onto an unfamiliar, relatively treeless, area immediately west of the old coast path route below Whitlands. 'Skeletons' of Carline Thistle which persist well into the winter suggested that there might be other interesting plants. Tree stumps indicated previous management, while the existing trees were up to 15 years old, suggesting management in about 1985. A Norman Barns map of the main tree species in the area at about that time pointed to him as the likely tree feller.

When Tom Sunderland took over from Albert Knott as Site Manager, he had two distinct advantages; he lived closer to the Reserve, but still some way away, and he did not have Albert's commitments to Dartmoor's woodland. However, he was soon to have responsibilities for Reserves over much of Dorset. Tom was ready to organise work in what became known as the Humble Glades as the area is close to Humble Point and Humble Pond. On older maps the pond appeared as Whitlands Cliff Pond while the area seaward of the glades was Humble Green. The first glade is on acid soil which explains the presence of Hard Fern and the abundance of young Birch, always an effective coloniser of open ground. Glade two starts with a small, bare, sandy area recently colonised by Ivy Bees and continues with a jumble of rocks projecting from the shallow soil where many of the bare patches indicate spots where Holm Oaks have been removed. Tom's co-worker, Rob Beard, has done much of the clearing and knowing the local geology, describes some of the complexities of the area. The previous numbering system now makes little sense for old glades two and three are now joined by another cleared area which was rapidly colonised by Marjoram, Blue Fleabane, Centaury and, inevitably, Hemp-agrimony.

Hard Fern on an acid soil

Above this, Rob describes how old glade three has large blocks of Beer Head Limestone which yield a rich and unusual fossil fauna. The best example of this unusual Cenomanian strata is the large block at the top of the bank which has a veneer of Upper Cenomanian beach material laminated onto its surface. A rich fossil assemblage is clearly visible in the bed, which is famed for its Ammonites, but it also contains gastropods, brachiopods, molluscs and echinoids. The bed has a very restricted range of exposure and is found only at Pinhay and Humble Point in the landslip.

Earlier, in February 2015, the then seven year old, Oliver Squire, joined a work party and his vivid account of the day featured in the next Newsletter. *"Twelve workers went down into the Undercliffs with Otter and Fuggles. There were four big cracks in the ground where the soil was slipping down the hill as we approached Humble Glades. Tom was the boss and he told us where to start the fires to burn everything that we cut down. I collected and cut sticks for the fire before climbing up to join Rob who was looking for fossils. I was lucky and soon found a tiny Shark's tooth, bits of shell and an Ammonite. When it was time for a break, Marjorie made tea and coffee and I had two biscuits. By now the fires were really hot so I was fetching wood from all over the place and while sorting out the fire, I picked up a log and dropped it on my foot".*

"At lunch time Fuggles stole Roger's glove and tried to steal lots of sandwiches. There was a disaster when Peter fell off a rock and spilled his drink which Otter tidied up. Donald and I then took the dogs for a little walk, and they splashed around in Humble Pond which stirred up lots of horrid smells. When we joined the others, Doug got his saw stuck in a big tree. Rob was trying to throw a rope around the branches to get the saw out when suddenly the tree came crashing down. By now the fires were really big, the dogs were wet and smelly and I thought it was time to head back to the cars. Tom thought we should let the fires die down a bit, so we had some more tea and biscuits, tidied up and started back along the little path that led us up to Whitlands".

The way back starts under the rocky glade where the fossils are. It merges into so called Glade 2 which Rob describes *"as running east/west and being underlain by a jumbled mass of Upper Greensand containing bands of siliceous chert from the middle part of the Greensand formation. This is also mixed up with blocks of Chalk of Lower Turonian age from the Holywell Nodular Chalk formation. The glade is dry, free draining and with calcareous soils which favour the many typical calcicoles. A bank of Foxmould at the eastern end is much favoured by Ivy Mining Bees".*

Continuing into the first glade, on our way back, its underlying Foxmould Sands make for unstable, waterlogged conditions prone to continual creeping movements which produce a landscape of gullies and seasonal wet flushes and pools. Like much of the Greensand, it tends to be decalcified by percolating rainwater forming damp, loose, sandy grey soils that become more neutral in character.

Had Oliver and I gone beyond Humble Pond, through what Rob calls the corridor, to reach Glade four, we would have found more massive blocks of Greensand and Cenomanian limestone protruding through the dry calcareous soils. Passing through more Holm Oak the 'path' reaches Glade five, large and free drying, with many blocks protruding through the bare soil. Rob has recently cleared more Holm Oak stumps and opened up the way to the final glade which, being underlain by Lower Turonian age Holywell chalk, is highly calcareous and has a fine flora of limited extent.

Yellow Rattle, as a partial parasite, can weaken aggressively growing grasses.

Back in 2010 or so, soon after work opening up the glades again, David Allen wrote a short account of Undercliff plants claiming that a walk through the Reserve rapidly reveals that here is our rainforest; a magnificent stretch of wooded cliffs that comprise some of the most unspoiled country in Southern England. The wonderful wilderness is itself dependent on the geology, geomorphology and climatic variation within the site. Under the heading 'Humble Hollows', as the glades were then known, he refers to a 'new' area of calcareous grassland and mentioned 16 of the typical species. Ten years later, in a note about an August visit intended to give my undisciplined observations a greater aura of authority, he again mentioned 16 species.

At the time of this visit, the glades were very dry and there was plenty of bare earth where tree stumps had been treated. Marjoram was, again, abundant and Wood Small-reed locally so. Tall Fescue, Hawkweed Oxtongue and the Everlasting-pea were frequent with the other nine species including Common and Lesser Centaury, occasional or rare. There were no longer any signs of the Yellow Rattle which had been abundant in a roped off square earlier in the year. Rattle seeds had been sown there so that the parasitic plant could weaken and reduce a particularly aggressive growing grass. As a plant introduced to do a particular job, it had been removed before it could set seed. Slightly beyond, was evidence of different work as trees had been felled to make it easier to reach the final glade which Phil Parr had started to clear some years earlier.

Before going into the Undercliff late in the summer of 2020, David and I had meandered among the flowers in the Allhusen's Lynch Meadow immediately above the Reserve. I feel its plant community must be similar to that which would have covered much of the Undercliffs before tree cover developed. Black Knapweed, Small Scabious, Dwarf Thistle and Burnet Saxifrage, apparently unrecorded in the NNR, were all frequent among the chalk grassland community. Earlier in the year, the abundance of Cowslip, rare in Devon, had, as always, been a glorious feature of these slopes. The Allhusens have recently given permissive access to the meadow as part of a clifftop path towards Pinhay House. Before reaching the house, the path goes down into the Undercliff, past some fine Beech trees. There are more below, around the coast path. They had all been planted by George's grandfather and been admired in 1950 when Major Allhusen met representatives of the Nature Conservancy, the Forestry Commission and the County Council to discuss the possibility of a National Nature Reserve between Axmouth Harbour and Ware.

That the glades were not only botanically rich, but also ideal for invertebrates, had been demonstrated by Phil Parr. As he became less mobile, it was good that Paul Butter showed an equal knowledge of butterflies and an even greater interest in moths. Towards the end of July 2016, Paul led a walk to what he described as *"a superb south facing xerothermic grassland"*. Gatekeepers and Meadow Browns were soon seen, and a Silver-washed Fritillary flashed by. It was good *"to come across first one, and then many, Oncocera semirubella, a very attractive micromoth restricted to the coasts of southern counties"* and often mentioned by Phil. There were plenty of another small micromoth, *Pyrausta aurata*, amongst the Marjoram, a larval food plant.

The second glade was *"a bit more rocky with quite a few limestone boulders prominent. There was more bare ground and shorter, herb-rich grassland where we soon came across second brood Dingy Skippers. While wondering whether it was too late for Marbled Whites, one was found being consumed by a Crab Spider. A little later, live ones were seen. On the slope I managed to net another micromoth which turned out to be the chalk-loving* Cochylis hybridella.

"The final glade had more large exposures of limestone with massive boulders scattered around and extensive bare ground. Three or four Small Purple-barred were seen near Milkwort, their *larval food plant. Even more excitingly, we saw several Chalk Carpet, brilliantly camouflaged, at rest on the bare limestone. A large patch of Hemp-agrimony attracted a number of Silver-washed Fritillaries and some of the group were fortunate to see a Dark-green Fritillary which settled for long enough to be identified".*

Two years later, in August 2018, Paul returned for another Butterfly Conservation event. In the woods below Whitlands a Rosy Footman and a Small Fan-footed Wave were disturbed. A Small Copper was seen in glade one, where a Silver-washed skulked in search of egg laying sites. In the next glade were second brood Dingy Skippers and a male Fritillary which zoomed around settling briefly on Hemp-agrimony or on Buddleia. A small white butterfly disappeared over the edge of the path, was it a Wood White? When a similar butterfly was netted, it was a male Wood White but, even better, one of the two Wood Whites seen on our return was a female. After lunch, glade three had more Dingy Skippers and speciality micromoths but the final, largest, open glade only produced abundant Common Blues and even more Skippers.

Silver-washed Fritillary

Rocky Glade 5 after management.

In January 2021, Paul gave me some details of his mothing experiences in the glades in 2017 when he had trapped 431 species including 80 nationally local, 15 nationally scarce (Nb), two nationally very scarce (Na), two Red Data Book species and seven recent colonists. It became clear that *"The glades host several species newly arrived in the UK, in particular, the Jersey Mocha, the Oak Rustic and the Sombre Brocade, which are all using the extensive Holm Oak.*

"There are also several long-term resident species that use some of the rarer plants with the Pyralid Moth, Cynaeda dentalis *on Viper's-*

bugloss, and *Ethmia dodecea* on Common Gromwell. The Mere Wainscot, which feeds exclusively on Wood Small-reed, has no recent Devon records away from the Undercliff. Wood Small-reed has been considered as an unwelcome aggressive grass but now some stands will be left uncut. The larvae overwinter in the lower parts of the stems, in a similar way to the RDB Morris's Wainscot, which does the same thing on Tall Fescue. Another less well known rarity is the Beech-green Carpet which has very few Devon records with mine from the glades being the first for ten years."

"There is no doubt that the creation and maintenance of clearings in the otherwise closed canopy woodland has created a great many habitat niches that would otherwise not exist and has, in turn, led to such a high biodiversity".

Paul also mentioned the Wood White for which the local coast is a national stronghold. "The continual landslips provide the bare ground necessary for pioneer plants like Bird's-foot Trefoil, the larval food plant, to flourish. Several of the newly created and extended glades now host the species along with Dingy Skipper which also feeds on the Trefoil. Continuing the butterfly theme, I have now had the first Devon record of Essex Skipper confirmed from the narrow corridor of uncut grassland on the coast path leading into the Undercliff.

"The species seems to have jumped from the Spittles to this small stretch of uncut grass, leapfrogging those areas where annual cutting and raking have created unsuitable habitat. The previous regime of short turf, annual cutting and arisings removal on Goat Island, the Plateau and Humble Glades, was in place to help some of the rarer flora, such as Autumn Lady's-tresses, Spring Gentian and Butterfly Orchid, but despite the good flora insect diversity was surprisingly low. A great many insects, especially overwintering larvae and pupae, need a longer sward to protect them from predators and frosts.

"We have now agreed on a three year rotation of cutting and raking which should allow niches for both ruderal flora and overwintering insects. Now that only a third of the area is cut at any one time, a September cut is not as detrimental".

The Essex Skipper provided the biggest butterfly news from the time of the first Covid lockdown which began on 24 March. By April, fine weather had set in and on the 21st came the first record of Wood Whites. Two days later he found good numbers of Green Hairstreaks,

Dingy Skippers and Wood Whites in the glades. As good weather continued the diminutive day-flying Small Yellow Underwing appeared and a mothing session on Haven Cliff, at the end of the month, turned up the scarce Ruddy Carpet and the RDB Morris's Wainscot, previously found by Phil. There were also 17 Cream-spot Tigers and a Small Elephant Hawk-moth. In mid-June, the same cliff produced the rare Pyralid Moth, *Cynaeda dentalis*. Pheromone lures at Ware brought in Red-tipped Clearwings, scarce in Devon. It certainly seems that the observant expert can still produce new records from the diverse Undercliff habitats.

Also, during lockdown, I had headed for Pinhay Warren with Woody, hoping to record any land movements there after a couple of years when life's events had limited Undercliff exploration. The 'path' onto the Warren was totally obscured by Brambles so we abandoned that project and lunched in the remains of East Cliff Cottage. When I sat down on old timber, I found that there was a nail protruding within half an inch of the bottom of my spine.

Soon after we had reported our failure to Tom Sunderland, he and Rob Beard cleared a way in, cut back dense Cherry Laurel around the cottage remains and banged the nail back into the wood. Later, we will see that Tom has plans for the Warren. My failures continued elsewhere when I got stuck in the mud of a tiny landslide at the bottom of the way from the Plateau to the Slabs. I was rescued by my son and

Distinctive "cow-stones" below Culverhole (1981)

The Sea-slater (*Ligia oceanica*) is common under the rocks. It emerges at night to feed on detritus.

Chapter 8

The warehouse at the mouth of the Axe

grandson, but more landslips obscured the usual way back up to the sheep-wash and the coast path.

A couple of months later, the same trio, plus granddaughter Freya, explored one of the two steep routes up from the old coast path under the Goat Island cliffs. At the top we looked at areas that have been partially cleared over the last three years, now colonised by Yellow Rattle, Yellow-wort and Marjoram, as well as a host of Thistle species. When Rob and Will had been clearing near the cliff edge a few days earlier, they had found two comatose Dormice and given them protection against the April chill. Not surprisingly, they had gone but we did find Slow Worms, a Common Lizard and emerging orchids on tiny scrub-free fragments of cliff edge. There were precipitous views over our way up which could be called the western ravine or goyle.

We later left the east end of Goat Island by the coast path steps before turning right along the route of the old path. A year earlier, despite landslips, it had been manageable but now, much overgrown, a series of deep drops and collapsed steps, explained my slow progress and when the point was reached for descent into Culverhole, to look at the ongoing erosion, I opted out and chose to continue along the old, but here solid, coast path.

A few days later an approach from Axmouth Bridge was less challenging. Past the Blue Marine Lobster Store and moored fishing boats beside the river, and the café, tackle shop and final house on the other side of the path, the base of the sandstone cliffs were covered in Willows and Sycamore. The house, now divided, had been created from a warehouse built in the late 1700s as a storage facility when the harbour was busy. Later, it was

Bare cliffs beyond the warehouse (1964)

160

reutilised by John Hallett, as mentioned earlier, but suffered with the arrival of the railway in 1868. In the early 1950s the building was damaged by fire and was then converted to the two houses whose porches and balconies were built around the old warehouse doors.

Beyond the boats and houses, the cliffs were stable for the moment but where recent falls had created open ground, Bird's-foot Trefoil and Violets had colonised. Elsewhere, Willows and Sycamores had stabilised the wet slopes. Further on, Sea Beet grew on the shingle beach at the base of the cliff of Triassic mudstones. Along the first quarter mile of cliff at least ten significant falls had created more opportunities for the Trefoil and for patches of Kidney Vetch. In places, Cretaceous rocks had fallen from higher up and plenty of ongoing movement was evident below the slip of 2014. Similar movement had changed the base of Culverhole and I had counted at least 50 falls on my way from the Harbour.

No doubt rain had contributed to these falls for the Pinhay data shows that while early spring 2021 was dry (only 5.9 mm in April), the previous seven months had been distinctly wet. The high rainfall continued for the rest of the "summer" with 344 mm in the next three months. Only once in the previous 150 years had May's rainfall exceeded 150 mm but in 2021, 181.5 mm fell. October, regularly wet, had 197 mm and as the year continued water, percolating through the cliff strata, probably caused as much of the extensive erosion as the waves breaking on the cliff base.

Sea Kale

David and Albert consult while botanising.

Buck's-horn Plantain

CHAPTER NINE
Plant and Animal Diversity from 2000 to 2015 and the Bioblitz of 2012

An early event in the new century had been the appearance of a slim volume *The Wild Flowers of the East Devon Coast* written by David J Allen. Discussion about the need for such a book had cropped up on a visit to the Undercliff when David and Alison Cox, from Devon Wildlife Trust, discussed the possibility of a pocket-sized guide to local coastal flowers. Sea Kale, Sea Beet, Portland Spurge, Rock Sea-spurrey and Rock Sea-lavender are among the 73 plants described and illustrated. Orchids and Gentians also feature prominently.

Later that year, I was invited by English Nature, or, more accurately, by Albert Knott, to collect information about the National Nature Reserve and to create a reference resource. By the end of the endlessly wet, cliff fall winter of 2001-2002, Nicky's daughter, Sarah, had converted my appalling scribble into 700 pages of better organised information about the history of the Reserve. The Philpot Museum in Lyme had been amazingly supportive as well as providing a wide range of period photographs. I also needed lots of up to date records of species from less familiar taxonomic groups, particularly flowerless plants, fungi and lichens.

As a result, Mark Pool and Roy Jefferies were invited to explore the Undercliff for a couple of days using our home in Combpyne as a base. Not having worked with bryophyte experts before, I was not entirely ready for the slow rate of progress, but they assured me that a lichenologist like Barbara Benfield would need much more time to explore the minutiae of the 'plants' found. Fallen trees in the Chasm impeded progress but every now and again a 'eureka' cry would indicate a new find, hand lenses and specimen packets would come out and after a quick spray of water, a probable identification would be made.

Chapter 9

Like the fungus enthusiasts who came a fortnight later, Mark and Roy liked to exchange ideas about tentative or firm judgments on identity and to test each other's memories as to whether a particular moss had been recorded in that 2 x 2 km-square tetrad. Mark was organising a survey of the whole of Devon supported by a devoted band of those competent to deal with moss and liverwort identification. With the small size of many species, microscopic examination is often needed, hence the sample packets and the need to keep examples from different tetrads separated until later examination.

A number of new species, outside the reserve but within the appropriate tetrad, were found on the edge of an arable field on the cliff top before we moved to a sunny lunch spot on Goat Island where a Clouded Yellow butterfly and overhead Siskins suggested early autumn. A good search at ground level, scraping delicately with knifepoint, produced more specimens to be filed away. Descending to the coast path and along through Hazel Grove, the large boulders were productive. Expected, and sometimes less familiar, species were ticked off with evident satisfaction. Using a lens, the beauty of some of the tiny mosses and their fruiting bodies was its own reward even if the names being exchanged meant little to me.

Next day, the damp acid areas as we approached Humble Pond produced additional species but the pond itself added nothing. I was delighted by a burst of Cetti's Warbler song. Mark and Roy then disappeared into a wet Alder flush and from their exclamations it was evident that discoveries were still being made. When they emerged, a navigation error brought us out onto unfamiliar ground which would later receive extensive management. I had to leave the moss men there as I was due to lead a walk in nearby Holyford Woods.

On 27 October, David brought Jeff Benn for a day long search for fungi. 38 species new to the Reserve were found. I took a very poor picture of Jeff photographing Hen of the Woods, *Grifola frondosa*, a Basidiomycete, and one of the new records. Jeff led an annual foray around the small Devon Wildlife Trust Reserve at Hawkswood. He also had kindly updated the nomenclature of Tom Wallace's list of the larger fungi which now, no doubt, needs a further update. Barbara Benfield was the next expert visitor, searching methodically and successfully for lichens as she would a few years later during the Bioblitz. On that 2011 occasion, she found 98 lichen species and a few days later, towards the western end of the Reserve, she had a list of 90. Both days produced new records and pushed the local list towards 200.

I had little idea at the time what a group of distinguished naturalists, all with close links to the Devonshire Association, had been lured to the Undercliff. Jeff was the County Recorder for fungi, Barbara had restarted the lichen report in the Transactions after a lichen-free 30 years, and Mark had reintroduced a bryophyte report after a similar gap. Roy had only recently moved to Devon but would soon be making the same sort of contribution. The county moth recorder, Roy McCormick, had set up productive moth traps in our garden in Combpyne and as close to the Undercliff as his power supplies would allow. David Allen, in his botanical, rather than fungal, role, would soon be describing finds in Humble Hollows where management was beginning. Roger Bristow was about to become the Devonshire Association's Butterfly Recorder and further in the future, Martin Drake would not only revolutionise the finding and recording of Diptera in Devon but find an almost unbelievable diversity of flies in the Undercliff.

Towards the other end of the experience spectrum, University students sometimes wanted to carry out a research project in the area as part of their degree course. Normally they were put off being told of the access problems and the difficult terrain but if they were sufficiently determined, it was a pleasure to help, and Albert gave any support he could. While on an MSc course in Environmental Conservation, Simeon Day investigated the possibility of introducing herbivores to Goat Island and the Plateau. He concluded that *"zoned, rotational grazing with primitive breeds of Goats or Sheep, in an extensive system should be introduced"* but also pointed out the difficulties of fencing and the daily attention that the animals would need.

In the same year, 2000, S. Luscombe completed a study into the spread of Holm Oak using aerial photographs from February 1957, in black and white, and from March 1990. Even without colour, the

Holm Oak on Whitlands Cliff based on aerial photography. 1957 above, 1990 below.

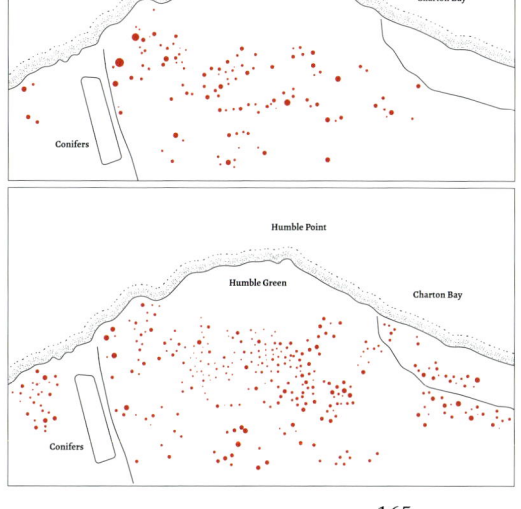

Chapter 9

The path to Charton Bay, shown in S. Lucombe's diagrams, photographed from the air in 1949.

leafy Oaks could be spotted from the air but counting them was rather more demanding. His conclusion was that *"the tree was spreading further and increasing in denseness every year"* and that the best management policy was to select the areas that most needed protection, such as the Ash/Field Maple woodland around Goat Island and the Chasm.

A 10,000 word dissertation in 2004 by Lorna Russell from Nottingham University was also concerned with Holm Oak. She asked whether there really was a need to manage it. Like most student projects, it was full of statistics and not surprisingly, concluded that the tree was spreading through the site at a rapid rate. When she thanked me and my dog, Branoc, for our help, my influence on her conclusion might have been obvious. If management was to continue, and if more glades were to be created, guaranteed funding was essential.

Lorna's work was acknowledged in a 2007 project in which Holm Oak, again, featured. Elizabeth Hankey studying Environmental Geoscience at Cardiff, chose the treacherous slopes slightly west of the Whitlands pumping station as the site of a transect. She chose it for *"the beautiful and demanding terrain"* and having helped her at times, I must certainly agree with her second adjective. The transect had to zig-zag due to ridges, fissures and crevices, as well as areas of impenetrable vegetation, and was illustrated in Chapter Four.

Branscombe church in the East Devon AONB. (See next page)

166

Following the Countryside and Rights of Way Act (1979), and between the first pair of projects and the second, a new East Devon AONB partnership was established. Its leader, Chris Woodruff, who had been working in the Forest of Bowland AONB, was supported by Nic Butler and Pete Youngman from the East Devon Countryside Service. Councillor Tony Reed was the first Chairman of the partnership and on 12 June he and Sir John Cave, representing the Country Land and Business Association on the partnership, welcomed 132 guests including 45 councillors and 14 members of the Partnership to Sidbury Manor for the launch of the new style AONB.

At much the same time, Albert wrote the fourth Management Plan for the years 2003-2008. Succession towards woodland had continued as *"the lack of recent (major) landslides, the absence of domestic stock, the reduction in the Rabbit population and minimal management had let natural succession lead towards the development of more mature woodland and a further reduction in species diversity.*

"Deer and Badger have most impact on the site. Key mammals include Dormouse" (12 nests would be found by contractors clearing around Humble Pond in 2007) *"and three species of shrew. The birds reflect the range of habitats present. Species of particular importance are Bullfinch, Song Thrush, Spotted Flycatcher and Marsh Tit. All have declined nationally and are UK Biodiversity Action Plan species."* (Three of these were, and are, doing very well in the woods but the Flycatcher barely holds on as a breeding species.) *"Reptiles and Amphibians are well represented by Adder, Grass-snake, Slow-worm, Common Lizard, Toad and Frog, as well as both Palmate and Smooth Newts"* (Harvest Mouse nests, some occupied, have since been found on Goat Island and Great-crested Newts have turned up among vegetation dredged from Humble Pond.)

Although most Undercliff trees were described as relatively young, they still supported some invertebrates associated with more mature woodland. The plan mentioned that Mike Edwards had found the hover-flies, *Fernandinea cuprea* and *Spegina cluniceps*, and the solitary

Bushy-tailed Dormouse in one of its less agile phases.

wasps, *Ectemnius ruficornis* and *Crossocerus dimidiatus*, among the woodland which he thought would benefit from the creation of more open glades with standing dead timber.

The Plan mentioned 19 species of insect which the 1987 Red Data Book 2 described as occurring somewhere among the soft coastal cliffs of the Sidmouth to West Bay SAC. Further on, there is another very similar list of invertebrates with the additions of *Priocnemis agilis*, a spider-hunting wasp, and the RDB3 Fly, *Herina oscillans*. Buglife describes it as a species not only of soft cliffs but also of coastal grassland, saltmarsh and fen.

A new feature of the plan, following the European designation of the coast as a Special Area of Conservation, was 18 full and part pages of tables explaining what 'favourable conditions' were for SAC features. Luckily, the assessment concluded that the local Management Plan and its specific prescriptions was consistent with the SAC and were not likely to have a significant effect upon them. That meant that no assessment was needed to ascertain whether the proposals might have an adverse effect on the integrity of the European site.

Six appendices filling 40 pages followed. A summary of the previous plan was first, then the citation of the 1955 declaration of the Reserve with a list of 29 operations likely to damage the special interests and thirdly, the reasons for recommending the area as a SAC. Appendix 4 was made up of the tables described in the previous paragraph, while 5 repeated the 11 page summary of the geological and geomorphological features of the reserve which had first appeared in the 1992 Management Plan. Finally, there was a summary of the geological sequence.

Two years later, Albert wrote an article for the magazine *Conservation Land Management*. In "Letting things slip on the Undercliffs" he wrote of the technique of photo-graphic monitoring of the changes which occur at the seaward face of the main landslips. Eight of these were photographed every spring so that changes could be detected by comparison with the previous year's images.

The Axmouth to Lyme Regis Undercliffs National Nature Reserve is on the move! Albert Knott explains the need to decide where, and where not, to intervene on a site that has been, and still is being, shaped by the natural processes of landslide and coastal erosion.

Right The Axmouth to Lyme Regis Undercliffs NNR slipping at Pinhay. *Kevin Page/English Nature*

Letting things slip on the Undercliffs.

These seaward movements are closely linked to those happening inland and above.

"The soft cliffs are both botanically rich and among the most important invertebrate sites in the country". Limited intervention, Albert mentioned, was a sustainable method of maintaining the rich mosaic of habitats. *"Parts of the Reserve fall into the sea every year but other sections are more stable and habitats such as grassland, are more at risk of being lost through natural succession"*. He then pointed out how management, particularly since 1993, had enhanced the nature conservation value of the grassland and described the threat of the spread of exotics, the role of the Countryside Service in maintaining the coast path, and the increasing publicity, both on and off site, leading up to the 50th anniversary of the NNR.

The anniversary gave Albert the chance to win Lottery money which enabled him to arrange for Stuart Line cruises to take children from primary schools in Seaton, Colyton and Axminster along the coast to Charton Bay where, as they landed, a splendid storyteller in appropriate regalia emerged from the undergrowth. Some of the Axminster children had never seen the sea before. The money also provided the opportunity for much clearing of *Rhododendron ponticum* from an area of acid soil below the Rousdon Estate. Other non-invasive species of Rhododendron planted by the Peeks were left.

Cover of "Exploring the Undercliffs" (2006).

Next year, the launch of my book *Exploring the Undercliffs* marked the end of the "understanding the Undercliff" project which had involved a series of guided walks and coastal cruises in the anniversary year. The launch was held at the Pilot Boat in Lyme Regis and presentations were given by Roger Critchard, described as an Undercliff resident of old, Tim Badman, World Heritage Site Team, Denys Brunsden, World Heritage Site Trust and myself.

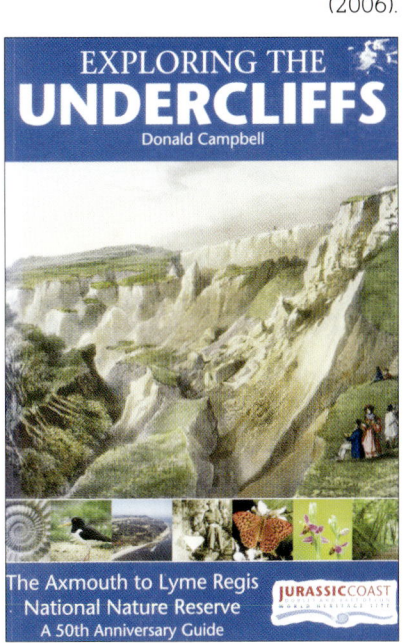

The invitation to the event, from Phil Collins, English Nature's Devon Area Manager, ended with a note on how the programme of anniversary events had helped the general public and local residents to enjoy and participate in nature conservation. *"This National Nature Reserve and surrounding coastline is a major attraction for tourists and an important asset for the local economy. We do hope that you will be able to join us in this celebration of the wonderful geology, geomorphology, wildlife, landscape and people described in this book"*

> **THE HERITAGE PLAYERS**
> present
>
> # WINEFRED
>
> A story of Smuggling, Love and Betrayal, and the Great Landslip of 1839
>
> **Directed by Penny Elsom**
>
> at Seaton Town Hall
>
> 26th – 30th July 2011, 7.30pm
> and Saturday 30th at 2.30pm
>
> From the novel by Sabine Baring-Gould
> adapted for the stage by John Seward and
> edited by Penny Elsom
>
> **Music arranged by Wren Music**

September 2005 also saw a visit to Seaton and the Chasm by the Sabine Baring-Gould Society who were celebrating the one hundredth anniversary of the book *Winefred, a Story of the Chalk Cliffs*. A follow up booklet gave a reinterpretation of a talk I gave with quotations from the novel. *"The whole Undercliff had a ruffled and chopped surface that was broken with ridges and depressed into basins"* came early on below an image of an almost treeless Chasm or gorge as it was apparently sometimes known. Slides showed *"the outlines of the Cliff, the features of the shore; reef and rubble, the line of torn seaweed and pounded shells and the site where Winefred had one of her periodic escapes from death"*.

In July 2006, David Allen brought a group of Devonshire Association members to Humble Glades where recent clearance had begun to remove Holm Oak and Buddleia. Some of the Oaks had evidently been cut down, some 15 years before, probably by Norman Barns, but there was abundant regrowth and plenty of young trees to be controlled. On the way down, David pointed out Spurge Laurel, and in the glades, showed members the chalk-loving plants like Yellow-wort, Lesser Centaury, Blue Fleabane and both Hairy and Pale St John's-wort. Grey Club-

The Jurassic Coast World Heritage Team
Left to right: Dr Sam Rose, Alex O'Dwyer, Sally King, Dr Anjana Ford, Caroline Pearce, Guy Kerr, Katie Thomson, Sam Scriven, Lesley Garlick

rush and Greater Pond-sedge were in Humble Pond and five plants of Sea Kale were found on the beach. Altogether the group identified 124 species but agreed that extensive management was still needed in the glades and in and around the pond.

The dry-stone walls in our garden in Combpyne also needed constant management but with or without that they were full of woodlice. After an AONB meeting in the month of David's walk, Sam Rose, Leader of the World Heritage Coast team and his former Leeds University tutor, Stephen Sutton, came for tea in the garden. Somehow the talk came round to woodlice, and it transpired that Stephen had written a book about them, part of a series on invertebrate types, and that I had bought a copy 32 years earlier. Stephen signed the book and then inscribed it welcoming *"the unanticipated opportunity to hail the incredible perception of the owner of this humble tome 'Per Isopoda ad astra'"*.

The walls in our garden were full of woodlice.

Two common species of woodlice - *Porcello scaber* and *Oniscus asellus*

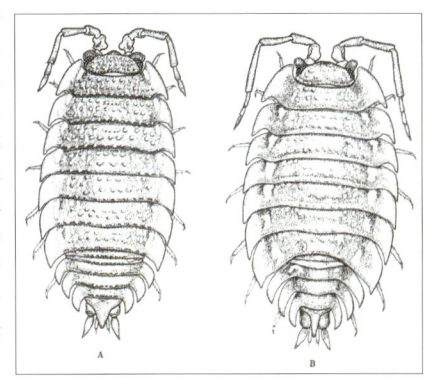

Two of those who could have written books on invertebrate types, and who did write a number of papers about them, were Mike Edwards and David Gibbs. Mike had started recording Undercliff invertebrates soon after Albert's effective management of the grasslands, which followed the work of Phil Page, had coincided with the increasing appreciation of the soft cliff habitat. In 2002, David, involved in a preliminary vegetation survey, *"took the opportunity to collect invertebrates on an opportunistic basis"*. He recorded 49 species, 12 of which had important conservation status. On Goat Island he found five nationally scarce species even without sweep netting. These were the Tawny Cockroach, the brilliant green Leaf Beetle, *Cryptocephalus aureoles*, a grey weevil, a large black Iris weevil, and a red and black spider hunting wasp, *Priochemis agilis*.

171

At Culverhole Point he spent *"a few minutes with a sweep net"* and was rewarded with seven nationally scarce or RDB species. These were the attractive but elusive Red-tipped Clearwing, a moth closely associated with species of willow, a ground beetle, a cranefly, three soldier flies, all small, like the black RDB3 fly, *Herina oscillans*, with black spotted wings. More extensive botanical surveys followed the next year but David, again, managed to find time to look for invertebrates. Five days were spent on the extensive soft cliff area of Pinhay Warren, three at Charton Bay, two at Humble Point and a single

Table 13

The Number of Species and of Red Data Book and Nationally Scarce Invertebrates at five Undercliff sites (David Gibbs)

Taxa	Culverhole Spp	RDB	Plateau Spp	RDB	Charton Bay Spp	RDB	Humble Point Spp	RDB	Pinhay Warren Spp	RDB
Araneae	2	1							2	1
Hemiptera	1	0			6	0	4	1	4	1
Lepidoptera	7	2	7	1	11	1	6	1	9	3
Coleoptera	1	0	1	1			9	3	13	5
Trichoptera							2	0		
Hymenoptera	7	1	1	0	16	6	13	2	20	3
Diptera	20	9	2	0	9	3	45	6	119	21
Totals:	**38**	**13**	**11**	**2**	**42**	**10**	**79**	**13**	**167**	**34**

day at Culverhole. He spent so little time on the Plateau that his finds there were not included in his species comparison histogram (opposite). Survey work involved searching flowers, potential food plants, dead wood and areas of bare soil for the more conspicuous species. Less obvious ones were collected by sweeping with a 14 inch diameter white bag through grass, herbaceous vegetation, over bare marshy areas and seepages down the cliff. Specimens were always taken for a later confirmation of the original identification.

Of the 292 species collected, 50 had official or pending nationally scarce status and 14 were RDB listed. Given more time, better weather and greater attention to common species, there would have been an even longer list. Of those 64 species, 35 were Diptera, 9 Coleoptera, 7 of both Lepidoptera and Hymenoptera, together with a couple of spiders and

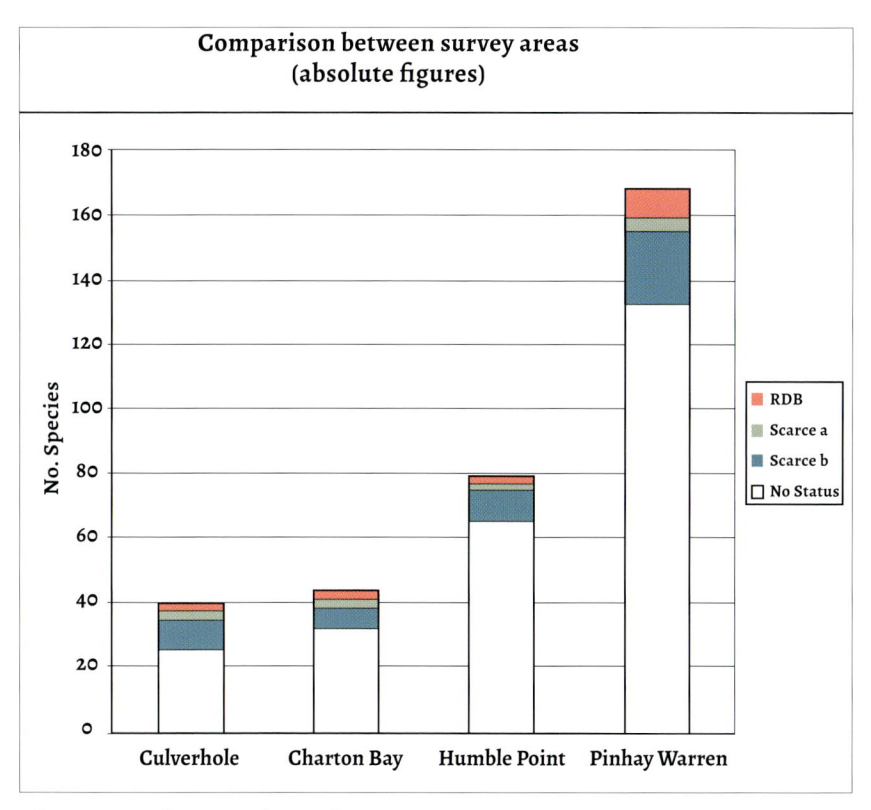

Comparison between survey areas (absolute figures)

Chart 3 - The number of RDB, scarce and other invertebrates at four soft cliff sites.

Caddis Flies. The single dragonfly, a Keeled Skimmer and the only cricket, Cepero's Ground-hopper were from Pinhay Warren. 19 species were new to Devon.

Phil Parr was, I think, the first to report the Ivy Bee in the Undercliff when he found some near Culverhole Point in 2003. The first in Britain had been reported from the West Dorset coast in 2001. Also in 2001, it was found 80 miles to the west, near East Prawle, and in 2002 it was at more sites including Branscombe. In 2006, he estimated that a colony at Dowlands was made up of at least 7,000 nests with each female making a single hole. The nectar and pollen, which she collects from Ivy flowers, is stored in the burrow as food for the larva which will emerge as an adult in the following August, perhaps surviving until the end of October.

Chapter 9

Talking to Phil in July 2001, just after seeing my first Wood White, he said that he had been suspicious about two white butterflies in the same area a few days before and had had six definite Wood Whites in Culverhole earlier in the year. He then spread his Undercliff maps on the kitchen table with all the potential good butterfly areas indicated by marker pen. Later that day he was to photograph a newly emerged Chalk-hill Blue on the Plateau, the first Devon record since 1981. Phil's search technique was to locate areas of Meadow Vetchling, Kidney Vetch and Bird's-foot Trefoil on the sunny cliff sites where the butterflies are likely to be.

The local Wood White stronghold was outside the Reserve in Dorset, but Phil had Devon records that year from Charton Bay and a soft cliff site east of the Plateau. Thomas and Lewington describe the butterfly as *"easy to recognise even in flight for it flaps its slender wings so slowly that their distinctive outlines are clearly visible, as are the males' black wing tips. The delicacy of the wings, their oval shape and its long slender body all distinguish this from other Whites"*.

In 2001, Phil recorded more than 230 Meadow Browns, mainly on Goat Island and the Plateau, 172 Gatekeepers, 169 Common Blues, 98 Dingy Skippers which, like the Blues, were mainly on the cliffs around Charton Bay, 28 Marbled Whites and 22 Brown Argus. His detailed records of butterflies and other wildlife were to continue for the next seventeen years and will be described later.

A new BTO survey started in November 2007 and was to continue until summer 2011. Timed counts involved four hours of recording all birds seen or heard within relevant tetrads. Not surprisingly, my survey started in the Undercliff going through the Chasm, over Goat Island and along to Rousdon before returning through Humble Glades, along the beach and up to the Plateau. That day, 15 species including Marsh Tit, Bullfinch and Jay were contacted more than ten times in SY 28 V and Z. As it turned out, Bullfinches ended up as the 12th most recorded bird, Marsh Tits the 17th and Jays even further down the list. Some birds are elusive at all times with Treecreepers only accounting for six records per thousand birds.

Bullfinches were particularly hard to find early in the breeding season when they remain very quiet. For this survey, coastal birds were included but on the exposed shore between Lyme Regis and Axmouth Harbour even Herring Gulls can be scarce. An occasional Red-throated Diver and a few Cormorants, Curlew and Oyster Catchers were usually around, and Peregrines, Ravens and gently chirping Rock Pipits are

Plant and Animal Diversity from 2000 to 2015 and the Bioblitz 2012

Red-throated Diver

Rock Pipit

Shelduck

Whimbrel

Common Sandpiper

Little Egret

Cormorant

Table 14

The Percentage of Contacts with the Most Recorded Birds early in the 2007-2011 Survey

Winter 2007-2008	
Wood Pigeon	13
Blackbird	8.1
Great Tit, Robin	6.4
Wren	6.0
Jackdaw	5.0
Long-tailed Tit	5.0
Blue Tit	4.6
Goldcrest	4.3
Chaffinch	4.2
Song Thrush	4.1
Bullfinch	2.7
Coal Tit	2.6
Carrion Crow	2.3
Nuthatch, Pheasant	2.1

Early Summer 2008	
Wren	12.4
Blackcap	9.1
Blackbird	9.0
Chiffchaff	8.6
Robin	8.2
Song Thrush	7.5
Great Tit	6.5
Wood Pigeon	6.3
Chaffinch	4.8
Blue Tit	3.9
Goldcrest	3.6
Coal Tit	2.2
Carrion Crow	2.1
Jackdaw	2.0
Long-tailed Tit, Nuthatch	1.8

good company. At other times, Fulmars, Shelduck and passing Whimbrel and Sandpiper add much to a unique stretch of wooded coastline. Later with a need to quantify bird records as part of a national coastal survey I found more Herring Gulls, 36 in all along the five miles (8 km) of shore below the Undercliff. I also counted 21 Curlew, 16 Oyster Catcher, 8 Cormorant, 5 Kittiwake, 4 Rock Pipits and 2 Herons as well as single Great Black-back and Black-headed Gulls, Little Egret and Peregrine. Hardly productive!

In both seasons, smaller numbers of Dunnock, Goldfinch, Marsh Tit and Treecreeper made up the top twenty.

In June 2015, I wrote an article on *Undercliff Butterflies* for the Devon Branch of Butterfly Conservation. It was almost entirely based on Phil Parr's records from 2001 to 2006 but to the totals for those years I have added his 2009 and 2010 records. The article also quoted his records of the flight periods of "interesting" species, so a number of their records were included in the text.

Table 15

The Most Recorded Butterfly Species 2001 – 2010 (Phil Parr)

Common Blue	1,013	Peak in August
Gatekeeper	788	11/07 to 17/08
Meadow Brown	744	20/05 to 12/09
Dingy Skipper	474	Increasingly in the glades
Small White	463	Very scattered
Painted Lady	332	Mainly 2009
Ringlet	216	28/06 to 02/08
Silver-washed Fritillary	187	Humble-Glades
Marbled White	175	Plateau favoured
Red Admiral	158	Flying into October
Speckled Wood	155	Along coast path
Wood White	129	Main site at Ware
Small Tortoiseshell	110	Scattered

Phil had seen a Small Blue on the Plateau in June 2005 and believed that the species might breed on the steep cliff below where there is plenty of Kidney Vetch. In 2010, he found it both at Culverhole, where Roger Laughton counted six, and at the base of the cliff, near the mouth of the Axe. In 2008, he reported that David Allen had seen a Purple Hairstreak that had come down from the Oak canopy on the edge of the first Humble Glade. In May 2010, Marjorie Waters had recorded five Grizzled Skippers below Goat Island and Brian Raine had also seen the Skipper as well as providing a very peculiar observation. *"I took a few pictures yesterday, 8 August 2008, of a nectaring Jersey Tiger on Hemp-agrimony and to my surprise in each one a dead White-letter Hairstreak could be clearly identified on the same flower, presumably victim of some nasty Crab-spider".* Numbers of migrant Painted Ladies and Clouded Yellows varied widely from year to year, while Walls and Holly Blues also showed wide fluctuations.

The *Millennium Butterfly Atlas* suggests a reason for the fluctuating Holly Blue populations, a reason that re-emphasises the significance of interactions between species. Holly Blue larvae are parasitised by a small wasp (*Listrodromus nycthemerus*) which can cause high mortality. As the parasite is specific to the butterfly, change in the population of the one is mirrored by a later change in the population of the other. The wasp, therefore, increases when the butterfly is common and

Chapter 9

Table 16

Flight Periods of "Interesting" Species of Regular Occurrence. One Wall was in March.

Species	April	May	June	July	Aug	Flight Periods
Dingy Skipper	65	72	59	199	59	16/04 – 14/06 08/07 – 24/08
Silver-washed Fritillary				68	73	17/07 – 27/08
Wood White	1	31	19	29	19	23/04 – 11/06 27/06 – 24/08
Brown Argus		5	6	8	41	29/05 – 24/06 14/07 – 25/08
Wall	11	12	1	5	26	17/03 – 05/06 19/07 – 25/08
Clouded Yellow		2			1	34 in Sept and 13 in Oct
Green Hairstreak	5	8	13			24/04 – 14/06

reduces the number of Holly Blues. This, in turn, makes life difficult for the wasp which decreases allowing a rise in butterfly numbers. With both Holly and Ivy playing important roles in the butterfly's lifecycle, and with Holly having a limited population in the Undercliff, the Blue is never common there.

The Green Hairstreak is another species that is not common in the Undercliff. Despite a wide range of potential food plants, it usually favours Gorse on Haven Cliff. One day when Phil was looking down from the golf course searching for possible ways down to Hairstreak haunts, he spotted a potential and accessible breeding site but also saw that someone was already there. Suddenly, a coastguard also appeared, followed by a helicopter whose pilot joined the coastguard. Phil acted innocent showing no interest in going down the cliff but watching as rescue followed. The man, who had clambered up but was unable to get down, was winched up into the helicopter. Apparently, he

Specialist rescue workers from the RSPCA look for Ben, a Yorkshire Terrier, who fell 250 ft.

Plant and Animal Diversity from 2000 to 2015 and the Bioblitz 2012

Lionel's rescue from The Elephant's Graveyard on Haven Cliff. It was in July 2001 after he and his mates had been to a local beer festival. He decided to scramble up to where they used to play as kids, but then having frightened himself realised going down was even more risky, so phoned the Coastguard.

had been at a beer festival in Branscombe reminiscing about games enjoyed on the cliff in childhood and had climbed up to enjoy his old playground.

Phil's reports ranged widely over the natural world with moths, beetles, crickets and fungi, as well as trees and flowers. Among his moths were the spectacular Scarlet and Jersey Tigers, Humming-bird Hawk, Oak Eggar and plenty of Six-spot and occasional Five-spot Burnet Moths, as well as less known species. Plenty of Silver Y's, the Undercliff regular *Onocera semirubella* and the distinctive Plume Moth (*Capperia pterophoridae*) were among these. Additional records came from Barry Henwood who used a pheromone lure to attract the wood-boring Six-belted Clearwing to show to Phil in Humble Glades. He also discovered Morris's Wainscot (*Photedes morrisii*) by torchlight near Culverhole Point. It is internationally rare and, in Britain, only found between Sidmouth and Black Ven where the Tall Fescue grows.

While some of the other insects like Green Tiger and Bloody-nosed Beetles and Rose Chafers are easy to identify, others demand a specialist. A nationally scarce Red Leaf Beetle was found by Marjorie Waters when botanising near Humble Point and identified as *Cryptocephalus bipunctatus* by Richard Wright. Among the fungi, Phil thought he had found Satan's (or the Devil's) Bolete in 2007 and two years later, the five inch cap of another Bolete on the Plateau was almost certainly *B. luridus*. The *B. satanus* identification was confirmed when David Allen found it again, under Holm Oak, in July 2009. He also identified an unfamiliar Ink Cap (*Lactarius acerrimus*) and the Zoned Tooth-fungus (*Hydnellum concrescens*) under Pedun–culate Oak on Whitlands Cliff. In 2010, the weird Red-cage Lattice-fungus (*Clathrus ruber*) turned up again, this time on Haven Cliff, and finds of the Warted Amanita (*A. strobiliformis*), from the beach road below Rousdon, and the spectacular Rustgill

Jersey Tiger

179

Chapter 9

Brown Argus

Silver-washed Fritillary

Green Hairstreak

Wall Brown

Plant and Animal Diversity from 2000 to 2015 and the Bioblitz 2012

The distinctive shape and pattern of the underside of Wood White wings as they nectar in Humble Glades

Small Blue

Dingy Skipper

(*Gymnopilus junonius*) near Humble Pond were confirmed by David Farley when Phil returned to Warwickshire.

The Management Plan for 2010 – 2015, written by Tom Sunderland, was shorter than previous ones. He began by commenting that *"The National Nature Reserve is one of the largest coastal landslip systems in Western Europe and the largest and most important landslip area on the British coast being the only Reserve in this country holding the entire range of important features characteristic of this type of environment. The geomorphology and the unstable nature of the terrain is by far the most important physical aspect of the NNR, having a significant bearing on every aspect of its management"*.

Tom described the site as an exceptional example of active land slumping that allows successional habitats to develop on bare ground following landslips and earth movements. In particular, woodland has developed as the dominant habitat across the Reserve in the last 150 years. Now its Flora is represented by the following main vegetation types (a) woodland, (b) grassland/fen communities, (c) calcareous scrub, (d) bare ground/sea cliffs pioneer habitat, (e) freshwater ponds and springs, as well as (f) littoral zone and foreshore. I have described some of these in detail, and others only in passing, but the shore's plant and animal life will be described in a later chapter.

A couple of years after his Management Plan, Tom wrote an article for *Newsletter 73*. *"Many of you will know that the AVDCS has been helping for many years with practical conservation work on the spectacular Undercliffs Reserve. Having been privileged to hold the position of Site Manager for Natural England over the last five years, I would like to take this opportunity to thank the volunteers for all their hard work in this time – the NNR is undoubtedly a better place for all your efforts. The western half of the Reserve is now largely free of non-natives and is considered to be one of, if not the, best example of unmanaged pristine ash woodland anywhere in the country. The management policy for the eastern half of the Reserve is not elimination of non-natives, but rather, acceptance with control where necessary"*.

The Reserve is only in 'favourable condition' if each of the invasive species occupies less than 5% of the total area. Buddleia and Cherry Laurel are two species which can approach this level of cover and hence are constantly being cut back with varying levels of success. Woodland below Pinhay is a particular Laurel haven but as Tom described, its destruction in this case revealed a lost ornamental garden, probably planted around 1850. Although landslips have altered the terrain over the years, the cutting back of the Laurel exposed walls, ponds and paths, as well as a range of exotic tree species including Coast Redwood, Maidenhair Tree, Chusan Palm, Bay, London Plane and a Cedar of Lebanon. His article then points to the *"delicious irony in the restoration and preservation of this exotic garden – having spent years trying to get rid of non-native species, we are now celebrating the discovery of a forgotten arboretum"*. However, as has happened many times, clearance without persistence tends to be temporary and the garden has been lost again.

In the same Newsletter, I wrote about a Bioblitz under the heading 'Squinancy-wort and Shredded Carrot Sponge'. I began *"what, you might ask, do King Alfred's cakes, a Piddock and a Noonday Fly have in common with Hairy Tare, Brussels Lace and Back-swimmers?"* A more logical start to describing the Undercliff Bioblitz is to refer to a report of the event during which 1,125 species were found on land in a day when another 150 were identified by Devon Seasearch diving in Charton Bay. Among their finds were Sea Mats, Sea Fans, seaweeds, Sea Hares, Sea Anemones, Sea Cucumbers, Sea Squirts and Grateloup's Fringe Weed. The range of taxonomic groups identified at sea will be summarised later.

The Bioblitz arose from a suggestion by Undercliff resident, George Neate, to Tom Sunderland. It was organised by the pair of them with help from Mike Lock and myself. Experts were gathered from around the county and elsewhere and during the day (and the night) some 600 Arthropods, 258 flowering plants and 67 mosses and liverworts were identified. Of the Arthropods, there were 186 two-winged flies and 252 moth species. In *Newsletter 82* of the Devon branch of Butterfly Conservation, Barry Henwood reported the find of two rare immigrant moths. The larvae of the Small Marbled were found on Fleabane flowers and the Plume, not recorded in Devon until that year, had suddenly turned up at four other county sites. The moth trap set up on the cliff edge of Goat Island attracted Devon's first Jersey Mocha (*Cyclophora ruficiliaria*) an immigrant first recognised in Jersey in 2002. A single Mere Wainscot (*Chartodes fluxa*), with few Devon records, was found on its locally abundant food plant, Wood Small-reed. Two coastal species, Northern Rustic (*Standfussiana lucernea*) and Hoary Footman (*Eilema carida*) were also noted.

A wide range of arthropods other than insects were identified. Toddy Cooper, who had caught 80 spider species and 10 harvestmen in pitfall traps in Humble Glades in 2003 / 2004, was disappointed with his 25 species but ten of them were new records. Most of the woodlice previously recorded turned up again. One find caused problems. When the name *Arcitalitrus dorrieni* appeared at the end of a list of miscellaneous species recorded during the day, few knew anything about the animal. It turned out that it was an Australian amphipod and an article on Google provided the background. R P O'Hanlon and T Bolger writing in Volume 24, Number 11, of the *Irish Naturalists Journal* (July 1994) described how terrestrial Amphipods have been recorded from coasts and volcanic islands in the Indo-Pacific, Caribbean and

Bank Vole

Two Bioblitz vertebrates

Common Lizard

Table 17 – Bioblitz Animals, A Summary
672 species of terrestrial animals

Phylum Annelida – **Class** Oligochaeta		
Earthworms	7	Identified by experts from Preston Mountford Field Centre.
Phylum Arthropoda – **Class** Arachnida		
Spiders	24	Despite 10 new records Toddy Cooper was disappointed.
Harvestmen	2	Two long-legged Opilionids
Mites	14	All gall formers identified by an unknown expert
Chilopods	15	11 Millipedes and 4 Centipedes.
Tick	1	*Ixodes ricinus* (probably quite a lot of them).
Class Crustacea		
Isopods	8	All the woodlice were given English names.
Amphipods	2	The Woodhopper and some fresh-water shrimps
Class Insecta		
Coleoptera	20	A poor total.
Dermaptera	1	The Common Earwig, *Forficula auricularia*
Dictyoptera	1	The Cockroach, *Ectobius pallidus*
Diptera	186	Martin Drake and Rob Wolton gave three of them English names.
Hemiptera	34	Thanks to whoever produced this impressive total of bugs, lice and plant lice.
Hymenoptera	14	Robin Williams was the expert on ants, bees and wasps.
Lepidoptera	20	Butterflies on Phil Parr's list.
Lepidoptera	252	Moths identified by Barry Henwood, John Randall and team.
Neuroptera	1	A lacewing, *Hemerobius micans*
Odonata	3	Two Hawkers and the Golden-ringed Dragonfly
Orthoptera	11	A good range of crickets and grasshoppers
Siphonaptera	1	Dog flea
Thysanura	2	Bristletails

Table 17 – Bioblitz Animals, A Summary [continued]

Phylum *Craniata*		
Aves	49	Including Gannet, Green Sandpiper. Peregrine and Raven
Mammalia	14	Badger, Bank Vole, Dormouse and a selection of Bats.
Reptilia	3	Adder, Common Lizard and Slow-worm.
Phylum *Mollusca* – **Class** *Gastropoda*		
Snails	2	A poor haul with not even a slug or a Banded snail.

Table 18

Plants and Their Former Allies (lichens and fungi)

Class *Bryophyta*		
Liverworts	12	Mark Pool and Nigel Pinhorne found and identified all the bryophytes and gave one of them an English name.
Mosses	55	
Class *Coniferophyta*		
Conifers	3	Norway Spruce, Scots Pine and Western Red Cedar.
Class *Filicinophyta*		
Ferns	10	Abundant Bracken and Hart's Tongue but few Black Spleenwort.
Class *Magnoliophyta*		
Flowering plants	258	Including 35 grasses and sedges and 28 trees and shrubs.
Class *Mycophytophyta*		
Lichens	98	An obsolete term for the cognoscenti; but still convenient for the rest of us.

Antipodean regions. The Wood-hopper in question is a native of Australia and the authors suggested that its presence in Britain was due to the passion of gardeners in the late 18th and early 19th centuries for the cultivation of exotic plants on which the wood-hopper had travelled.

For many, bats are as unfamiliar and as unloved as a Wood-hopper, but developments in technology are making it easier to identify the species

Adder

and appreciate something of their remarkable abilities. Fiona Matthews from the University of Exeter has done much to promote interest in the group and to increase our knowledge of them. With their Anabat detectors she, Kimmo Evans and others spent the night monitoring all bats on the way down into the Undercliff and on the coast path above Charton Bay. Not surprisingly the Common Pipistrelle with 75 'passes' was most recorded followed by the Soprano Pipistrelle with 44 'passes' and Myotis species (Whiskered and Brandt's Bats) with seven. Five Long-eared and a single Noctule were also recorded at the same time as the moth enthusiasts were collecting on Goat Island. A summary of all records from the 'blitz' is in Tables 17-18, on the two previous pages.

The pathetic total of only two fungi was due to the early season, the dry weather and the absence of a mycologist.

Green Sandpiper

Water cascade on unstable Haven Cliff. Its "squidgy" stream edges are a haven for flies and the cliff was the only site where Martin Drake found all eleven of the commonest species.

CHAPTER TEN
Flies, Moths and Another Look at Trees

The Bioblitz report described how *"the springs, seepages and squidgy stream edges, where the soil remains wet all year, are a haven for flies and no doubt Martin will add many more to his list"*. After reading that his efforts in a parish biodiversity audit had converted the Chardstock list from 550 to around 850, I asked him for an update on his Undercliffs' records. By 2019, his total there had moved from 186 to 591. Regardless of flies, Martin Drake had joined most of the local work parties over the last ten years, always busy destroying Brambles, or raking up cut grass, but most active perhaps when grappling with invasive Cherry Laurel. He is often partially visible in a tangle of the dense shrub before working his way down until he can get at the many roots, developed over time as the plant spreads.

In the 2004 annual report on the activities of the entomology section of the Devonshire Association, Phil Cook outlined Martin's career. For his PhD he studied the midges of a chalk stream. He then became an entomologist with the Freshwater Biological Association at Windermere and went on to be one of English Nature's specialists before setting up his own consultancy. Phil's report also explained how the Biodiversity Action Plan process had not been effective for flies which had been omitted from the 1981 Wildlife and Countryside Act. In 1987, Red Data Books included 830 endangered, vulnerable or rare fly species; by 1991, this had increased to 1,500 and in the next year SAC's (Special Areas of Conservation) were designated for listed species but not yet for flies.

In Devon, a Fly Group was established in 2013, an initiative of the Dipterist's Forum led largely by Rob Wolton who had found 830 fly species in a single one of his farm hedges. When field meetings were arranged, Martin was much involved and soon distribution maps no longer petered out at the Dorset and Somerset borders.

In the Undercliffs, Martin's RDB records included nine Grade two and Grade three Crane-fly species and two Picture-wings. The only Grade one fly, where a species is in danger of extinction if causal factors continue unabated, is *Helius hispanicus* which has been mentioned elsewhere when Alan Stubbs found that its habitat on Haven Cliff had been *"clobbered"*. RDB2 species, vulnerable ones, are likely to move into Category one *"if causal factors remain in operation"* while Category three species have small populations that are at risk. The status list is now in the process of further change and is, as Martin points out *"a horrid mixture of old Red Data List statuses and recent IUCN statuses. I give precedence to the IUCN status, but anything that has IUCN status has to be RDB in old money"*. Table 19, with old money terminology, includes records from six sites along the Undercliff.

Martin is writing the RES handbook on Dolichopodid flies, two of which are found nowhere in Britain except Black Ven. The Undercliff does not have so many extreme rarities, but it does have 86 species, about a quarter of the British list. As Devon has about 215 species (70% of the British list), there are probably more to be found in the Undercliff. Two flies from other families top the county's special species list. One, *Idiocera sexguttata*, *"must be the prettiest Cranefly with 'designer grey' pattern on the thorax, as well as spotted wings. It is on my list of RDB flies but Lipara similis, the Cigar Gall-Wasp, is not. It has another Devon site, Branscombe, where it makes narrow galls in the reeds at the point where the stream comes out onto the beach"*.

By way of contrast, Table 20 gives details of the status and locations of some of the commonest flies.

In reply to a request for more information about Martin and the flies he has studied, I got fascinating details of both, as well as the comment *"It is a shame that the county boundary splits the Undercliff from Black Ven and the Spittles as the Dorset end is easier to get to, so I have several rather interesting records from the wrong county from your point of view.*

"From a natural history perspective, I find the cliffs along the local coast among the most extraordinary places for flies. Maybe the Norfolk Fens beat them, but that is probably because I have done considerably more survey work there than on the cliffs. What strikes me most is that, when out in the field, one sees a range of apparently common species, but back under the microscope they

Table 19
Red Data Book flies in the Undercliff up to 2017

Crane-flies	Status	Where Found
Gonomyia abbreviata	RDB 2	2 Whitlands, 1 Ware
Idiocera sexguttata	RDB 2	1 Haven Cliff
Paradelphomyia ecalcarata	RDB 2	1 Ware
Arctoconopa melampodia	RDB 3	4 Ware
Gonomyia abscondita	RDB 3	1 Ware, 1 Culverhole
Gonomyia conoviensis	RDB 3	3 Ware
Gonomyia tenella	RDB 3	1 Ware
Orimarga juvenilis	RDB 3	1 Ware, 1 Haven Cliff
Orimarga virgo	RDB 3	4 Haven, 3 Pinhay, 1 Culverhole
Picture-wing Flies		
Campiglossa malaria	RDB 1	1 Plateau
Myopites inulaedyssentericae	RDB 3	1 Ware

Table 20
The Most Recorded Fly Species are all "Common" or "Local" except *Herina oscillans* which is "Near Threatened"

Family	Species	Records	Where Found
Pointed-wing Fly	*Lonchoptera lutea*	25	All sites except Goat Island
Picture-wing	*Herina longistylata*	20	All sites
Crane-fly	*Tipula lateralis*	19	12 from Harbour / Haven
Snipe-fly	*Chrysopilus asiliformis*	19	All except Plateau / Dowlands
Snail-killing Fly	*Hydromya dorsalis*	19	All except Plateau / Pinhay
Soldier-fly	*Oxycera pygmaea*	19	17 from Harbour / Haven
Shore-fly	*Hydrilla maura*	18	All except Goat / Culverhole
Shore-fly	*Parydra coarctata*	18	All except Goat / Plateau
Long-legged Fly	*Liancalus virens*	17	11 Harbour / Haven
Long-legged Fly	*Raphium brevicone*	17	All sites, with 7 Harbour / Haven
Picture-wing	*Herina oscillans*	15	11 from Harbour / Haven

are the rarer versions and the common ones do not live there – too hot, too little litter, or whatever.

"I am most interested in the wetland families, targeting water margins, streams and seepages. Even though the Undercliff seepages occupy a tiny area, they support most of the interest, especially those with a pronounced base rich influence. Alan Stubbs has, rightly, always made the point about the importance of a base rich habitat. To prove that he could tell what base rich status a sample of Crane-flies came from. I once handed him some to identify but only after I had dropped them and muddled them up. He came back telling me that the sample was ecologically impossible. So, flies really do have quite specific habitat requirements.

"The Devon Fly Group is making a big difference to our understanding of the county's fauna. We have missed a target we set ourselves, to publish a centenary fly list in 2020, to update Colonel Yerbury's lists in 1919 and 1920 when he had about 800 species. We have almost 3,300 now. Going back to my own history, my mother told me that I was keeping Woolly-bear caterpillars when I was four. At University I had already amassed an untidy collection of moths and beetles so asked if I could do a group that I did not know about – flies. Never looked back!"

In Chapter eight, Paul Butter's enthusiasm for, and knowledge of, butterflies and moths was mentioned, along with some of his records from Humble Glades.

As far as butterflies are concerned, the Wood White has a national stronghold between Charmouth and Salcombe Regis. The frequent landslips provide the bare ground necessary for pioneer plants like Bird's-foot Trefoil, its larval food plant, to flourish. *"Several of the newly created and extended glades now host the species along with Dingy Skipper which also feeds on Bird's-foot Trefoil. Years of annual cutting and raking on Goat Island and the Plateau have created suitable conditions for some of the rarer flora but removed the cover that a longer sward would provide. That would protect insect larvae and pupae from predators and frosts. We have now agreed on a three year rotation of cutting and raking which should allow niches for both ruderal flora and overwintering insects. Wood White chrysalids are attached to dead stems and scrub which is typically found at the grassland margins, the edge habitat being very important. The earlier regime of cutting all the grasslands was invariably carried out in September and removed all late flowers*

Flies, Moths and Another Look at Trees

Small Elephant Hawkmoth

Cynaeda dentalis at Haven Cliffs

Below left - Red-tipped Clearwing

Below right - Four Spotted Footman

Chapter 10

with their nectar source at one swoop at a time of the year when many migrant insects arrive in need of refuelling. Now that only a third of the area is cut at any one time, a September cut is not detrimental."

Paul had had the first Devon record of Essex Skipper confirmed. It differs from the Small Skipper in having ink-black tipped antennae and, in the male, a short black line on the forewing rather than a long curved one. The record came from the narrow corridor of uncut grassland on the coast path approach to Goat Island during the first Covid lockdown of 2020 which had begun on 24 March. A month later, fine weather had set in, and very good numbers of Green Hairstreaks, Dingy Skippers and Wood Whites were flying in the glades. As good weather continued, the diminutive day-flying Small Yellow Underwing appeared, and a moth searching session at the end of the month turned up the scarce Ruddy Carpet and the RDB Morris's Wainscot previously recorded by Phil. There were also 17 Cream-spot Tigers and a Small Elephant Hawk-moth. In mid-June, the same cliff produced the rare Pyralid Moth, *Cynaeda dentalis*. It certainly seems that the observant expert can still produce new records from the diverse Undercliff habitats.

A large Hybrid Lime by the coast path below Pinhay House.

I was not looking for new records when I started on two small projects in 2021, trying to age grasslands through estimating the size of anthills and measuring tree trunks' circumferences at shoulder height to compare with Hamish Archibald's measurements in 1955. The criteria for judging size was fine, except that many of the largest trees started branching very low and others were on steep slopes where access to the whole trunk was difficult; in those cases, the circumference could be calculated from the easier to measure diameter. At the Lyme Regis end of the Reserve, Sycamores are dominant, initially densely packed, except for those in the hedge bank along the south side of the approach to Underhill Farm. It was evident that the majority of trees in the woodland were the descendants of a few veterans which

Flies, Moths and Another Look at Trees

Above - Past Ravine Pond.

Right - Impressive trees beside the path leading up to Whitlands.

Below - Maintaining the sheep-wash.

pre-dated the woodland nature of the Undercliff. One of these old trees, double trunked and measured near ground level, was recorded as 524cm in circumference with many others over 300cm.

Further on, Sycamores remained dominant but much more scattered among dense Bramble, Bracken, Blackthorn and Old-man's Beard. One Ash measured 360cm and two Beeches were almost as large. Further west, by the clearing below Pinhay House, large hybrid Limes had a girth of around 460cm. They had featured, as had a neighbouring Beech, in the Great Trees of East Devon. This project was set up by the District Council, East Devon AONB and English Nature, and funded by the Heritage Lottery. It ran from 2005 to 2008 under the guidance of Kate Tobin who recruited Jon Stokes, Director of the Tree Council, Archie Miles, tree photographer and author, and myself as the judges in selecting 50 trees from the many put forward by local residents who had photographed and written about the virtues of their favourites. The nominated Beech has since fallen but a photograph of it with James Chubb sitting on a branch, gives an indication of how large it was. Higher up the steep slope towards Pinhay, and along the permissive path towards Whitlands, are a number of straight trunked Beeches with circumferences up to 340cm. Nearby is the idiosyncratic and marvellously positioned observation post built by John Ames so that he could look out for, and report, on Russian vessels during the Crimean War (see p.14). The flint work and roof were recently renovated by English Nature, but a nearly neglected sheep-wash is not in such condition. Below the permissive path are the largest Pines in the Reserve.

James Chubb in the large Beech which has since fallen

The remains of West Cliff cottage today.

Back on the coast path, past Ravine Pond, there are plenty of large Holm Oaks on the way down to the old pumping station. More impressive trees grow as the path climbs towards the ruin of West Cliff Cottage and up to Whitlands, while on the left are a Turkey Oak, 420cm, and a Sweet Chestnut of 460cm. At the top, in the middle of the path, a Holm Oak, also around 460cm, is not quite as large as another slightly up towards Whitlands. By an old wall, a Horse Chestnut, perhaps the only one in the Reserve, measured 340cm, a suspect figure because of its double trunk. Leaving the coast path to follow the route of an earlier way toward the sea, one passes the remains of field and orchard boundaries associated with the old cottage and a plantation of Norway Spruce, now in disarray with fallen and falling trees of around 220cm. Soon, Birch, Hard Fern and Horsetails indicate the acid soils derived from the 1840 landslip above. Near the sea, towards Humble Point, a group of young Alders are growing while larger Birches reach a surprising 185cm.

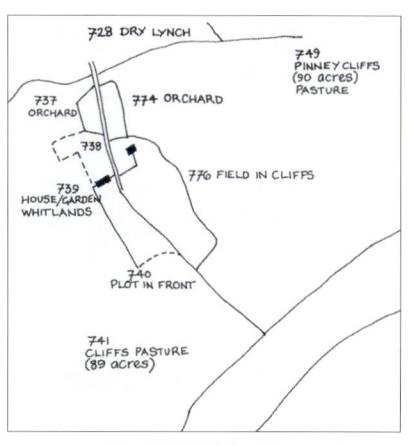

The 1840 Tithe Map, redrawn.

There are yet more Holm Oaks above Humble Pond with the low spreading branches of some suggesting pioneers. The very low branches of a Turkey Oak, a little higher up, and of a distinctly old Hornbeam with some dead branches hanging below the bank on which they both grow, suggest the possibility of an old hedge. Climbing back up to the coast path, through Bluebells and chaos, past a 297cm Field Maple which is perhaps the largest of its species apart from old hedgerow examples in the Chasm. Near the highest point of the coast path, below an impressive length of inland cliff, is a Yew, one of the very few in the Reserve, with hard to measure trunks estimated as 280cm.

Many of the largest trees were on steep slopes.

Later, I came along the path from the other direction. After leaving the managed grassland of Goat Island and passing a number of self-coppiced Hazels each with several dead or dying 'trunks', surrounded by a ring of healthy new growth, I went down the series of steps to join the old coast path. The Hazels were around 200cm and a number of Hollies, trees rather than bushes, had similar girths. There are more of these beside the steps. Along the path, many tall Sycamores survive to grow but the Ash trees may soon be lost to disease. Further on, more coppiced Hazels are among the boulders which fell from the Chasm as it opened up in 1839. The boulders destroyed two cottages. A hundred years later, Home Guard volunteers looked out to sea for signs of German invasion with boy runners, no doubt scouts, ready to take any news of sightings back to Lyme Regis. After another few years, Landslip Cottage was finally abandoned, and trees have grown destroying any view out to sea. Above the remains of the cottage and of the older sheep-wash, an Ash was measured by a group of children from a school in Kenton visiting Seaton in 2000 and learning as they explored. They decided to measure the tree by linking arms around it and found that it was five boy or six girl units in circumference. All had walked out from their base and were to return there along the beach, exploring as they went. Back in Seaton, they immediately began playing a variety of energetic games. The tree fell shortly afterwards and decayed rapidly as the fungi that had killed it multiplied. By 2021, its annual rings were hard to count but there were at least 90 so that it predated the War and the Home Guard volunteers and was significantly older than the trees which now obscure any view of the sea.

Where there are no trees, as on parts of Goat Island and the Plateau, there are often old anthills and where trees or scrub have been cleared, there may be smaller, younger ones. Some comments about ants and anthills described in *The Soil* were quoted earlier, but there was no reference to a formula that had been proposed by T J King for calculating their age. On Goat Island different areas have anthills of differing size and frequency. On the western side of the main grassland there is one every 10m^2 and using the formula, their size suggests that they are around 100 years old. Below the area of scrub to the east there is only a smaller heap every 25m^2 with an age of perhaps 25 to 45. Phil Page introduced the cutting back of scrub in that area in 1988 so the theory seems to hold. On Goat Island east only a small area of shallow, orchid rich, soil was free of scrub before the late 1990s. There, the heaps are spaced out at about one every 10m^2, as elsewhere, but some are larger than those to the west suggesting an age of at least 150 years which almost takes us back to the time of the landslip.

Flies, Moths and Another Look at Trees

Abundant ant hills cover grassland on parts of Goat Island.

Deep rooted Rock Rose and Wild Thyme flourish on the well-drained ant hills.

Yellow Meadow Ant with a pupa, often called an ant's egg.

"Bringing a slurry of grey mud onto the beach"

CHAPTER ELEVEN
Along the Shore

On 14 March 2021, towards the end of a year of isolation and Coronavirus, I meandered along the beach west of Lyme. Surprisingly, a Cetti's Warbler produced a burst of song, seemingly from under one of the chalets between the carpark and the beach. Behind the beach huts beyond the chalets, the sloping cliffs were dominated by flowering Gorse and well leaved Holm Oak but further on the cliffs steepened and instead of beach huts their base was covered in a slurry of grey mud, reed roots and tree branches. With the tide out people and their dogs had plenty of space. The beach, initially sandy with scattered stones, changed to layered limestone where colonies of Sabellaria worms covered the rocks. There were holes created by Piddocks and plenty of Keel and Tube Worms attached to stones and brown seaweeds. Higher up the seaweeds were green and higher still, fossil hunters hammered rocks in hope. The dog walkers preferred the easier conditions nearer the sea. Further on, the cliffs became steep and layered like the beach itself but though there was little water cascading down them, plenty of rock had fallen recently to form mounds at the base in which there were trees fallen from Pinhay Warren above. Superficially, it was much as it had been twenty years earlier, even if the Oystercatchers must have been the descendants of those described by John Fowles when he wrote that *"the sea sparkled, Curlews cried. A flock of Oystercatchers, black and white and coral red, flew on ahead of him, harbingers of his passage"*.

In March 2001, it had been foot and mouth disease and not Covid that had led to the closure of the coast path and increased the

Oyster Catcher in flight

Chapter 11

The normally clear stratification was hidden by recent falls.

numbers on the beach but on a grey day there had only been a few people exercising their dogs. The top of the beach, below the cliffs, was littered with bits of Holly, Birch and Horsetail roots, as well as the remains of Reeds from the sodden soils above, which were on the move after a long, wet winter. As the cliffs replaced the more gradual slopes nearer Lyme, there were further signs of winter subsidence as a great mass of detritus had accumulated at the base of the cliff.

Having rounded this fall the striations reappeared only to be obscured again by another fall, this time of lumps of black rock, which had created a new gulley down which the water ran, bringing a grey slurry of mud onto the beach. Along much of the length of the Pinhay Cliffs the falls were smaller but the continual sound, like falling porridge, explained why the normally clear stratification was hidden except where streams kept the Lias clean. An occasional Coltsfoot flowered on firmer ground.

A typical landslip west of Pinhay.

Parts of this cliff had been falling throughout the winter of 2000-2001, much of the rest of the local coast had also shown its instability, but movement was most evident at this east end. Falls high on the inland cliff below Ware and lower down at Pinhay Warren had changed the land-scape. Chalk cliffs west of Pinhay Bay had collapsed and further movement at the west end of Charton Bay made access up to the coast path far more difficult. The falls at Whitlands, which moved the pumping station, and from the face of Goat Island, were small compared to the 600 metre length of cliff which collapsed to lower the coast path as described earlier. Finally, turning inland at the mouth of the Axe, parts of Haven Cliff once again blocked the riverside path.

A couple of years before these events, members of the Axe Vale Conservation Society had had their first trip out to sea to look for birds from Harry May's boat *Predator*. Simon Tidswell was crucial for the success of the day as he had prepared buckets of fish remnants to be tossed overboard to attract the birds. Inevitably, these offerings tempted the Herring Gulls which followed the boat a long way out to sea where, eventually, the white wing flash of a Great Skua was the first of the day's excitements. Ten miles out the number of Gannets began to increase, and Simon's fish tempted some of the 300 following a trawler, where fish were being gutted, to join us. They followed and dived within feet of *Predator* but suddenly attention switched to something much smaller as a Storm Petrel pattered over the waves not far from the boat. It was fast and elusive but eventually everybody had good views. Other petrels were seen and up to 14 Great Skuas but on that occasion, no Arctics or Pomarines, both of which were seen the following year. A distant Sooty Shearwater was another highlight while Fulmars, Kittiwakes and Terns added to the day's success.

Shortly before this birding boat trip, I had first met Colin Dawes, joining him and a gathering of children for an expedition to look for fossils along the beach. He first set us a task in rock identification, searching the beach for some of the distinctive types. Soon a pile of 'Beef', grey, slightly dappled rocks made up of layers of calcite crystals had been collected and next, a similar heap of Chert, described as "blocky" and made up of yellow or brown silica, had been put together. He then asked us to collect flints, which were also siliceous, but knobbly and often with split white surfaces, and Chalk which, like the flints, had been moved east by longshore drift. It had fallen from the cliffs further west where the Upper and Middle Cretaceous beds overlie Chert and Greensand.

After this rock sampling it was time to find fossils. As well as Clams and Scallops, a 'nest' of Brachiopods and a marvellously preserved Micraster Sea Urchin were found before we reached the ammonite

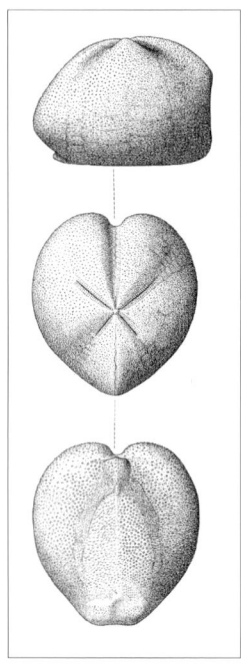

Chapter 11

Micraster coranguinum

pavement. In *Fossil Hunting Around Lyme Regis* (2003) Colin described how *"the first in the sequence of 'zone' ammonites lies under the beach below the base of the Blue Lias but the distinctive Schlotheimia angulata does crop up above sea level, on a limited stretch of beach where the Blue Lias arches up. It, and its relatives, give way to giant ammonites which range from the base of Monmouth Beach towards the top of the Lias. This range includes the Bucklandi Zone, named after Arietites bucklandi. The next zone ammonite is a species of 'Arnie' (Arnioceras semicostatum) which puts in an appearance at the top of the Blue Lias and extends into the lower part of the Shales with Beef"*.

The subjects of his next book, in 2005, were birds, many of which were to be found well beyond the local beaches. In the preface to *Bird Watching Where Dorset Meets Devon*, Colin explains how his booklet *"doubles up as a sort of celebration following the establishment of superb bird watching facilities in the Axe Estuary"*, facilities which have been greatly extended and continue to develop 17 years later.

Early in this book he features Lyme's conspicuous gulls, commenting on their apparent greed and messy nesting habits, but also referring to an acclaimed book about their behaviour *The Herring Gulls' World* by Niko Tinbergen in the *New Naturalist* series. Colin then had a page on the somewhat less conspicuous Purple Sandpipers *"which breed within the Arctic Circle before flying south with about a dozen of them overwintering around Lyme Regis. The flock normally keeps together, probing around Broad Ledge when the tide is out, and roosting about the Cobb when the tide is in. The bird is no bigger than a Starling. Its feet are yellow. So too is the base of*

Purple Sandpipers below the Cobb

Along the Shore

Herring Gulls follow *Predator* out to sea.

Gannet

Kittiwakes breed east of Exmouth

Herring Gull

Great Skua

Great Black-backed Gull

Common Tern

Fulmar

205

Chapter 11

its bill which fades out to a black tip. Its overall plumage is mottled with shades of brown and dark grey except for whitish underparts. The overwintering birds of Lyme Regis hardly live up to the common name of the species. They blend in remarkably well with rocks and seaweeds. They are tame in the sense that they will carry on feeding until you get within a couple of metres of them. Get any closer and they will hurry along in single file before suddenly taking to the air in a V-shaped formation".

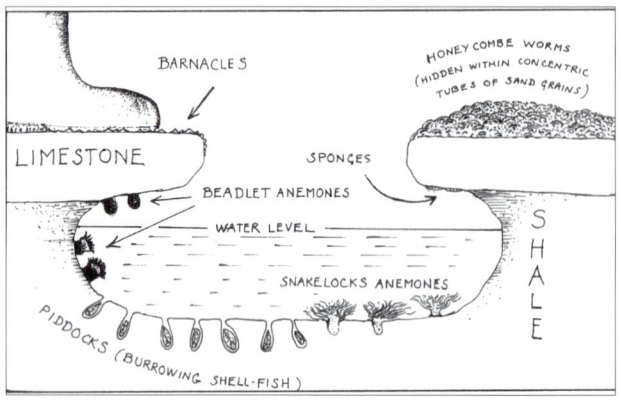

A rock pool in the Blue Lias

A year after the bird book, Colin returned to the local beaches with *Rockpooling Around Lyme Regis* which had a fine Tompot Blenny on its cover. Early in the book he shows a rock pool created in the Lias by the sea's erosion of limestone and the activity of boring Piddocks. He advises readers that the cleanest rock pools, with less sediment making it easier to observe the life, are in the reefs to the east of the town.

Colin 'expatiates' - one of his favourite words

The great grandfather of another well-known Lyme figure, Ken Gollop, was vaguely Piddock-like when he broke into the Blue Lias as he worked on the stone boats collecting limestone and clay to make bricks. I talked with Ken in May 2001; he was Chairman of the Trustees of the Philpot Museum and a fourth generation West Dorset resident who, in a varied life linked to the sea, had had much to do with boats and fishing. The earliest date he mentioned was 1853 when there was a railway from a cement works on the beach and a Blondin tower, part of an aerial bucket transport system which survived until 1920. Remains of the railway are still there today. Ken's great grandfather had been killed in 1893 when a cliff collapsed on him as he was knocking the rock into handleable sizes. In the 1880s the cliffs were a source of the clay and of the Lias limestone which were both collected by boat from east of the Cobb as, if one was working to the west, the boats had to be rowed around the wall of the Cobb.

When the stone boats finished, his grandfather went into fishing and then into pleasure boats. Later, when his son, Ken's father, was old enough, they bought a motorboat. I had come to ask about fishing, although always happy to hear about other aspects of life around the Undercliff. Ken had brought along a map on which he had marked areas for different fishing activities. Pots, mainly for Crabs, were out on the rocky coasts from the Cobb to Seven-rock Point, from Pinhay Bay to Humble Point, and then along much of the coast east and west of Culverhole Point. In the past there had been no market for Spider-crabs, so Ken and others had ended up eating them while the edible Crabs were sold.

The coast was used extensively by boats from Lyme, Axmouth and Beer catching Crabs, Lobsters and Prawns. *"In Pinhay Bay you have got an area of pure sand and they used to trawl there but they do not do it much now as the boats are*

Stoneboats by the Cobb, ca. 1910

Unloading from a stoneboat

The Cement Works and brick stores ca. 1890

Chapter 11

far too big. With a 24 foot boat you could make two or three circuits eight or nine times a year. There is another area in Seaton Bay, a big sandy area, as you have got to have clear ground; a hundred yards outside, you have had it, you would hit something. The Seaton and Beer men would trawl around these but outside that area you have rough ground, which is no good, not rocky enough for pots, so the only thing done now, and probably in the past, was netting. You set the nets overnight, a wall of netting, and haul it in the next morning. Now they are after Bass but in the past Whiting and Pollock would have been good".

I mentioned that there were still plenty of pots out there so Ken continued *"in the old days fishermen would only work 20 pots, 25-30 at the most, they were old willow pots which only retained a quarter of the catch, the rest got away. Now there are a couple of hundred more efficient pots strung along below the Undercliff".*

Pebble gathering, for grinding cosmetics and toothpaste, was another activity and when the organiser, the owner of Seatown beach, could not get his tractor down to Charton Bay after an Undercliff landslip, Ken and his brother were hired to pick the stones off the beach where they were stacked. *"We would load up the boat with the tide coming in, so that we could float off and get back to Seatown to unload before high water. Once there was an order for pure white stones, for a film set, so he had all these chaps on casual, paid by the sackful".*

When two trawlers and the associated fish shop became too much work, the brothers went their separate ways with Ken carrying on with boat trips, even when he had bought the aquarium on the Cobb. *"I would also take barbecue parties down of an evening, to Charton Bay and run a little shuttle service; Taunton Young Conservatives would come down twice a year. Another thing I forgot about, until the big freeze of 1963, from February through to Easter, one of our boats, and another one, were the first in the water after Christmas to go prawning. There was a big patch of mud, just off the Cobb, west prawn ground, and a big patch of mud just east, east prawn ground. We would have 45, and Victor would have 45, small willow* pots and we would go out first light, haul the pots, to be back by 9.30am, boil up the fish, get it up to the station by 1.00pm, they would be collected off the train by Billingsgate merchants and be on sale in London restaurants by 7 o'clock that evening. That was February to April. In the cold weather of 1963, we tried, and we tried, and we did not catch a single Prawn. One of those pots, 18 inches in diameter, would have had 100 big Prawns in, that would be a good little earner for us. £40 worth of Prawns a day for two men working half a day was more than you would get in an accountant's office! Then the big freeze came, and we put out Prawn pots and Lobster pots and did not catch a darned thing until a Lobster in April. The sea temperature was right down, it was bloody cold, we would bait the pots, haul them up the next day and there would be nothing, not even the little Winkle shells you get. The bait would be in exactly the same condition. In the end, by the end of March, we put out a few Crab pots; in those days you only caught Crabs and Lobsters from Easter to September. They now catch all the year round; how modern fishing sustains itself I do not know because of the efficiency. Nobody has ever caught Prawns on that same scale since".

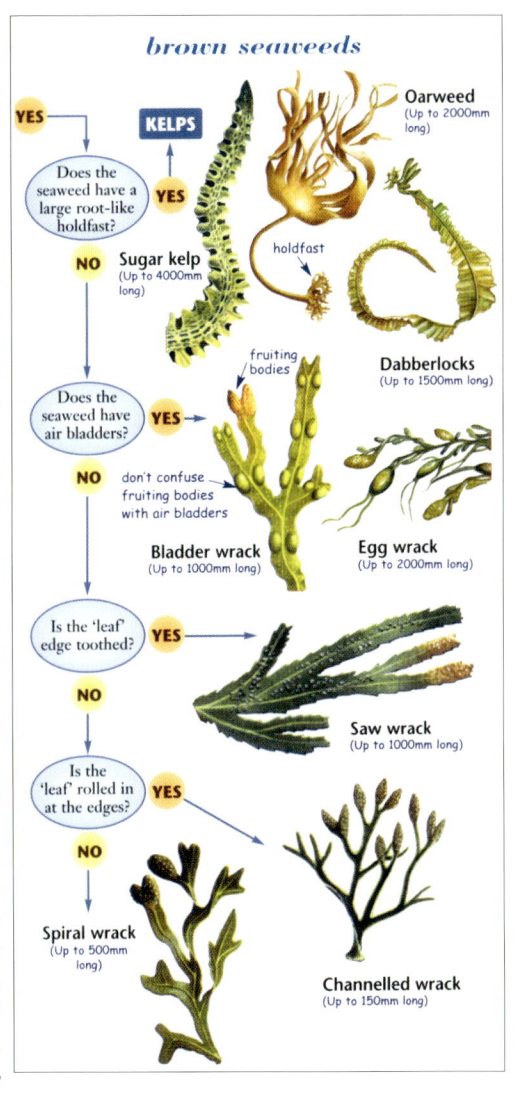

brown seaweeds

- **KELPS**
  - Does the seaweed have a large root-like holdfast? **YES**
    - **Oarweed** (Up to 2000mm long)
    - **Sugar kelp** (Up to 4000mm long) — holdfast
  - **NO**
    - Does the seaweed have air bladders? **YES**
      - don't confuse fruiting bodies with air bladders
      - **Bladder wrack** (Up to 1000mm long)
      - **Dabberlocks** (Up to 1500mm long) — fruiting bodies
      - **Egg wrack** (Up to 2000mm long)
    - **NO**
      - Is the 'leaf' edge toothed? **YES**
        - **Saw wrack** (Up to 1000mm long)
      - **NO**
        - Is the 'leaf' rolled in at the edges? **YES**
          - **Spiral wrack** (Up to 500mm long)
        - **NO**
          - **Channelled wrack** (Up to 150mm long)

His final memory, not very precise with regards to date, *"was in the mid-1960s, I think, when there was a big earth movement down there by the Humble rocks. I remember going down that morning, we were potting along there, and we left one day, and we came down the next morning and there was this wall of boulders up here*

Chapter 11

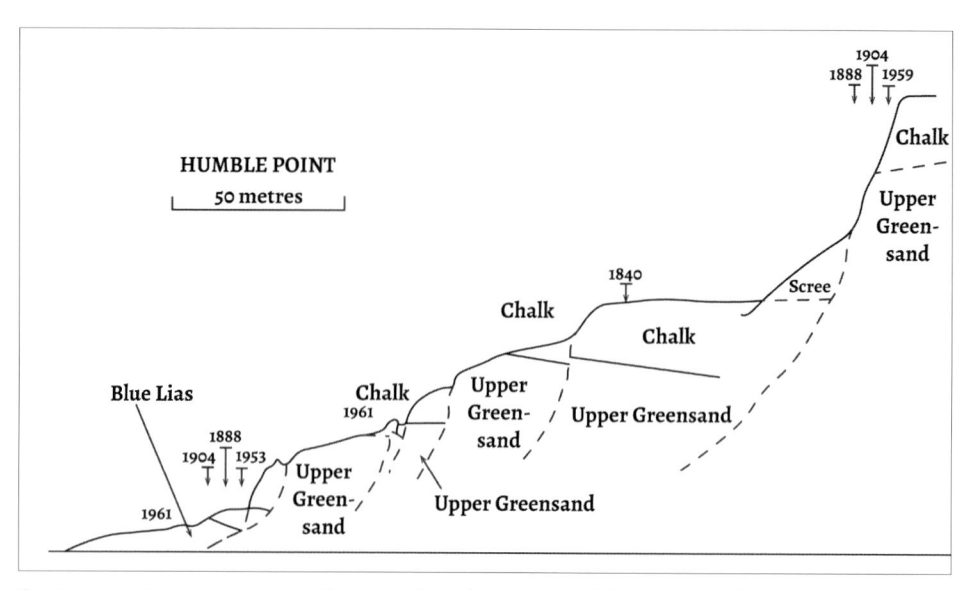

Section showing past positions of certain slope facets derived from map evidence (John Pitts 1981).

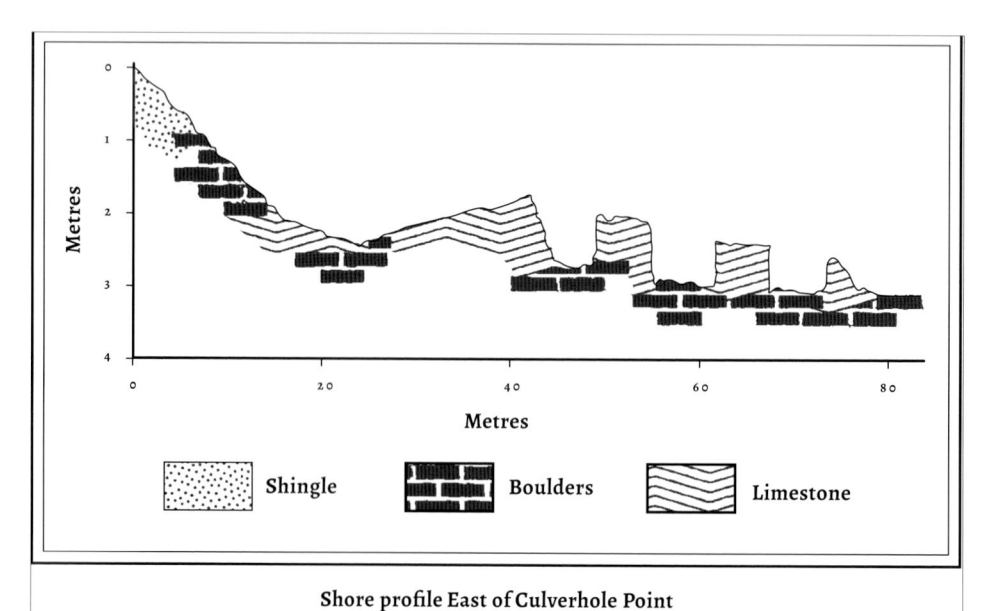

Shore profile East of Culverhole Point

Table 21

The Zones at three Undercliff sites and the number of associated plants and animals found at times of spring tides between 23-28 May and 22-27 June 1995

Pinhay East – 316904	Plants	Animals
Mid eulittoral limestone platform	9	15
Bladder Wrack platform	22	11
Lower eulittoral Serrated Wrack limestone	33	22
Platform pools with boulders and sand	47	21
Humble Rocks – 306898		
Supralittoral fringe	9	2
Boulders at top of shore	5	11
Serrated Wrack zone	11	9
Lower eulittoral rock platform	49	25
Pools and overhangs	29	8
Culverhole Point – 273893		
Supralittoral fringe	6	4
Bedrock lumps in upper-mid eulittoral	11	16
Mid eulittoral boulders	19	23
Lower eulittoral with broken bed rock	23	21

and they should have been down at sea level. It was a very, very eerie feeling. If you had been in the fog and came into that area, and hit that hundred yards, you would wonder where the hell you were".

Humble Rocks, east of Humble Point, was one of seven sites investigated by Patrick Armitage who described them in the *Proceedings of the Dorset Natural History and Archaeological Society*, Volume 91 (1969). His work on freshwater sites was mentioned in chapter seven and here, again, the influence of fresh water, whether trickles or more substantial flows as at the Gusher, was of particular interest. Although his title involved the littoral fauna, he inevitably considered some of the seaweeds and as their zonation, below Humble rocks was, and is, fairly typical I will quote him in some detail. Refer to diagram at left - below the shingle was a zone of Blue Lias boulders around two foot in diameter, above a slightly land dipped Rhaetic limestone ledge which in places had been eroded to form pits lined with boulders. Flat Wrack

Chapter 11

(*Fucus spiralis*) grew on top of these boulders, some 30 metres from the start of the transect. From 31-46 metres, the limestone was densely covered with Bladder Wrack (*F. vesiculosus*) but the pit between 46 and 50 metres was more varied with Serrated (*F. serratus*) and Flat Wracks (*F. spiralis*) as well as Bladder Wrack which also covered the next flat stretch. The second pit 54-60 metres along the transect was dominated by Serrated Wrack which also occurred in the final pit where it was joined by *Laminaria saccharina*. This species, which may have fronds 20 feet long, is more often found below the low water mark. Elsewhere, on this Sabellaria reef transect, *Corallina* sp. and Pepper Dulse (*Laurencia pinnatifida*) a flattened red seaweed, occurred 26-35 metres below high water.

Along his seven transects, Patrick found over a hundred animal species with 65 (28 crustaceans, 14 molluscs and 13 Annelid worms among them) occurring at multiple sites, 32 less frequently, and 11 at locations between transects. The most common of these 11 were Blennies, Piddocks, the barnacle, *Chthamalus stellatus*, and the Sea Hare, *Aplysia punctata*, one of the Sea slugs. Tom Wallace made use of these Armitage records including 12 species of fish of which only Eels and the low tide rock pool species Blenny and Rock Goby were common.

Steep, unstable Culverhole Point is above a boulder-strewn beach which was the third site investigated by Ambios consultants.

Many years later, in 1995, Ambios Environmental Consultants studied 17 sites between East Lulworth and Orcombe rocks as part of the Lyme Bay Environmental Study. Three of the sites, Pinhay East, Humble Rocks and Culverhole Point, were below the Undercliff. On many of the littoral fringe areas the cobbles and boulders were too mobile to support the usual upper shore lichens and algae but had the green *Enteromorpha* instead. Channeled Wrack (*Pelvetia canaliculata*) was found only at Humble Rocks where the Small Periwinkle (*Littorina neritoides*), an indicator of exposed conditions, also occurred. Pools at Pinhay East were richer in algae than elsewhere with at least six genera which were not on Norman Barn's 1989 list which despite 5 green, 19 brown and 23 red species, *was "merely the result of regular beach patrolling"*.

In the upper and middle zones, the dominant barnacle was *Chthamalus montagui*, often with *Semibalanus balanoides* and *Chthamalus stellatus* (all of which have six plates even if fused ones obscure the count) with the occasional *Elminius modestus* (four plates). For myself, barnacle identification is for others, and I sympathise with those who used to classify them as molluscs rather than the crustaceans that they are. Barnacles are very firmly attached to the rocks, but it is the remains of crustacean antennae that provide the cement that attaches them. The feeding barnacle combs the water with six pairs of "feathered" appendages or cirri which are derived from the walking legs of more typical crustaceans. In *The Seashore* C M Yonge wrote that on an exposed shore there might be a thousand million barnacles per kilometre liberating a million million larvae in a year!

Sea Hare - do the broad upper pair of head tentacles look like hare's ears?

Three limpet species on the higher shore were not safely identified as their pallial tentacles were not examined. Colonies of the polychaete worm, *Sabellaria alveolata*, are found in sandy patches as at Humble Rocks and to the east but not at Culverhole. Another polychaete, *Polydora*, is often unnoticed except when its two antennae wave vigorously. Animals further down the beach, in the sublittoral, including sponges, hydroids, anemones, polyzoans and ascidians, all intolerant of desiccation, were not investigated.

Another summary of records from the tideline was sent to Albert Knott in 1998. It came from Roger Covey of English Nature's Devon, Cornwall and Isles of Scilly team and included 12 animals and 13 seaweeds. None of those recordings had gone below the low tide line so Devon Seasearch's investigations as part of the 'Bioblitz' were particularly valuable. They recorded 150 species in three one hour dives in Charton Bay. Because sponges and Cnidarians are unfamiliar to many, I have listed records of these in full, with added comments from other marine authorities for some of them.

As sponges require a firm surface on which to attach, they only occur on rocky shores. They are filter feeders using flagella to draw water in through pores in the body wall, engulfing suspended matter and ejecting the water through a different, large, opening, the osculum. 22 species were found in the three dives. Some of the imaginative names

Table 22
Sponges Recorded in three dives during the 'Bioblitz'.

Common Sponges in Charton Bay
Orange encrusting species found at all sites.
Cliona celata. The Boring Sponge bores into limestone along the coast.
Sponges that are frequently found
Amphilectus fucorum. Shredded-carrot Sponges are encrusting and flexible.
Tethya citrina. The Golf-ball Sponge has a short stem under a lumpy ball.
Suberites carnosus may be reminiscent of a brain.
Stilligera stuposa is another without an English name.
Occasional
Dysidea fragilis. Its appropriate name is the Goosebump Sponge.
Suberites ficus. The Sea Orange or Sulphur Sponge is archetypal in structure.
Raspailia ramosa. Colonies of Chocolate-finger Sponges suggest small dense bushes.
Polymastia penicillus. Chimney Sponges are often part buried in sediment.
Haliclona viscosa. Volcano Sponge. Its water outlets look like mini volcanoes.
Pachymatisma johnstonia. The Elephant-hide Sponge has hard smooth surfaces.
Rare
Grey encrusting Sponge found on dive 2 only.
Hemimicale columella. Crater Sponge.
Axinella dissimilis. Yellow Staghorn with branches to maximise its contact with food bearing currents.
Haliclona oculata. Dives one and three.
Ciocalypta penicillus. Tapered Chimney Sponge.
Dercitus bucklandi. Black-tar Sponge.
Guancha lacunosa.
Thymosia guernei. Mashed-potato Sponge.
Present
Stelligera rigida.

Table 23

Anemones found by Devon Seasearch in July 2011 with descriptive comments from Paul Naylor (*Great British Marine Animals*) and Peter Hayward (*Seashore* NN94)

Cnidarians common in Charton Bay
Sertularella gayi, a colonial species with stalkless polyps.
Halecium sp. Herring-bone Hydroid, made up of branching colonies.
Obelia geniculata, with zig-zag stems and polyps enclosed in a bell-shaped cup.
Epizoanthus couchii, Sandy Creeplet.
Isozoanthus sulcatus. Ginger Tiny or Peppercorn Anemone.
Aiptasia mutabilis. The Trumpet Anemone with kelp-like brown colouring.
Those frequent in the Bay
Nemertesia antenina. Antenna Hydroid or Sea Beard with yellow brown tufts.
N. ramosa. Branched-antenna Hydroid with branding and sub-branching stems.
Actinothoe sphyrodeta. The White-striped Anemone is small and common.
Anemonia viridis. Snakelocks Anemone with symbiotic green photosynthetic algae.
Occasional
Hydralmania falcata.
Alcyonium digitatum. Dead-man's Fingers, a soft coral in varied colours.
Cerianthus lloydii. Burrowing Anemone. The only tube anemone with no basal disc.
Calliactis parasitica. The Parasitic Anemone which shares a Whelk shell with a Hermit Crab.
Rare but a much valued species
Eunicella verrucosa, Pink Sea-fan. A horny coral usually found below 10 metres.
Present
Edwardia sp.
Sagartia elegans. Elegant Anemone is small and often vividly coloured.
Metridium senile. Plumose Anemone has many small tentacles.
Caryophyllia smithii. Devonshire Cup-coral. A scarce shallow water stony coral.

Chapter 11

Shredded Carrot Sponge

Trumpet Anemone

White Striped Anemone

Volcano Sponge

Dead Mens' Fingers

Sunset Coral

of these species were mentioned in the context of the Bioblitz in an earlier chapter. Barrett and Yonge in their *Pocket Guide to the Seashore* comment that *"identifying sponges is a specialised study and largely depends on microscopic inspection. The least show of doubt seldom disappears"*.

Cnidarians, more familiar to many as Coelenterates, include Jellyfish, Sea Anemones, Hydroids and Corals. They have a body wall with only two cell layers, separated by the jelly-like mesoglea, and have unique eversible nematocysts, barbed stinging threads, often with powerful poisons. Although some of the group exist as single organisms, most of them are colonial with the archetypal situation involving sessile polyps linked by a stolon which is the basis of the colony. To avoid more names, I will not list the 14 species of Sea-squirts or the 9 Bryozoans, but only mention that on the dives they found 26 mollusc species and 18 crustaceans including 2 Porcelain and 4 Spider Crabs. They also recorded 15 fishes, 7 polychaete worms, 4 echinoderms and single Phoronid and Platyhelminth worms. The fishes included Dragonet, Greater Pipefish, Hooknose, Trigger-fish and both Corkwing and Ballan Wrasse.

Snakelocks Anemone

Back in 1992, local divers had begun to voice concerns about the state of the various reefs in Lyme Bay, particularly Beer Home Ground, Lanes Ground and Sawtooth Ledges. The reefs of Beer Home Ground are part of the sunken coastline which was flooded after the ice age. The reason for a new survey was the number of inshore trawl fishermen who, after finding difficulties catching Whiting, Cod, Bass and Lemon Sole, were turning to scalloping using dredges attached to a very heavy bar which did immense damage to the sea bottom and its associated life. In June 2001, I interviewed Chris Davis, the Lyme Bay Reef Project Officer, who had started surveying the reefs in 1998, and in the course of 88 dives across the bay investigated the cobble and large boulder communities which, below 20 metres, were largely dominated by animal communities.

Chris mentioned particular concern about Pink Sea Fans and the rare Sunset Coral. The Sea Fan had a substantial population throughout the reefs but was limited to South West coasts. The coral was found on Sawtooth Ledges but only two other sites in the UK. He went on to tell me that *"there are lots of Bryozoans and Hydroids and the overall diversity makes it one of the most interesting animal dominated ecosystems in the South West. This is due to a combination of depth, substrate and not too much tidal movement but enough to bring food to the animals"*.

Following the 1992 survey and a failure to consult sufficiently with fishermen, Chris had come in at a time when there was already visible damage. *"The concerns were, yes, it is still there, some reefs are just about okay, species are surviving in pockets, but if fishing level increased at all, we doubt if the reefs would be sustained. The reefs off Exmouth were once diverse and animal dominated but are now just gravel over very, very large areas.*

"We still want to close areas as the only true way to protect wildlife, but we must give something back to the fishermen. We have looked at a number of ways, such as artificial reefs and recreating habitats. We could also enhance the scallop population by putting in seed stock as has been done successfully in Jersey, Northern France, Scotland and even Portland Harbour or improve lobster stock and habitat as they have at Padstow.

Peacock Worms

"If this is not successful, we will push for a statutory closed area, as around Lundy with its marine nature reserve. The only species with statutory protection at the moment is the Sea Fan, protected under the Wildlife and Countryside Act, but a stray scalloper is not going to be much concerned about that. We would be careful to try to identify areas that were not too important to the trawlers and scallopers, trying to bring them into discussions.

"We know that the impact of scallop fishing is to remove animals and habitat, as can be seen on a video survey we have produced. It is like cutting down and then ploughing ancient woodland. There are certainly parallels between the ecosystems, such as the longevity of species; a big Sea Fan may be from 50-70 years old with a growth of 2-3cm a year.

"Local fishermen definitely act like wardens with regards to large nomadic boats; it is these opportunistic boats which are the great problem. We must try to encourage the fishermen to look after the local environment. How that is to be done is the difficulty and ensuring their future; but we are really encouraged by their attitude".

Pink Sea Fan

During Seaton Marine week in 2001, Chris talked about this wealth of diversity to be found in Lyme Bay. Starting in shallow water, he showed slides of the Tompot Blenny, the vicious Velvet Swimming Crab, the Snakelocks Anemone and the Devonshire Cup Coral found down to 80 metres. Deeper down, among the Kelp, life was influenced by the dwindling light and the nature of the currents. There were thousands of Peacock Worms and Trumpet Anemones 100 metres off the Undercliff, and the Yellow Boring Sponge was one of 80-odd sponge species in Lyme Bay. The Dahlia Anemone, 8-9cm in diameter, carpeted the bottom of gullies. Where it was totally dark, on the reefs he had described in the interview, were seven of the nine UK Corals, a high diversity of Sea Squirts and, where the current flowed, abundant Plumose Anemones. The rare Sunset Coral, the same diameter as a 50p piece, had been found a mile off Lyme Regis and the Pink Sea Fan, a colonial coral, was a protected species.

In 2010, Richard White, speaking at Devon Wildlife Trust's AGM, told how the 2009 Marine and Coastal Areas Act stated that the Government had a duty to designate Marine Protected Areas, MPAs. Two years later, plans for 127 conservation zones were unveiled, but late in 2012 only 31 of these were approved. Since then, by stages, 60 more zones have been established including the Axe Estuary. As well as the Trust, the Marine Institute of Plymouth University has been researching in Lyme Bay. As mentioned in my *Rocks and Wildlife Around the Axe* the 206km² MPA established in 2008 and described by Dr Emma Sheehan in 2013 had seen significant increases in both Dead Men's Fingers and King Scallops, as well as an increase in the Sea Squirt, *Phallusia mammillata*

Campaigning by Devon Wildlife Trust and conservation partners led to Lyme Bay reefs being designated as part of England's largest Marine Protected Area covering 215 sq. km of seabed. This protection benefits the wildlife-rich cold water reefs from damage by bottom-trawled fishing gear.

and an improvement in the population of Ross Coral. The number of Pink Sea Fan were still fluctuating.

Populations of Pink Sea Fans have now increased eightfold. By 2013 there were significant increases in Dead Men's Fingers. The sedentary species now provided habitats for fish including Ballan Wrasse and shellfish like Edible Crabs.

Now the marine biologists from Plymouth have been monitoring populations for more than ten years and the Sea Fans and King Scallops have increased eightfold within the Lyme Bay Reserve. Corals, anemones and sponges now provided habitats for fish and shellfish so that the ecosystem has benefitted as much as the fishermen. It would be good if these productive areas of Lyme Bay could join the Undercliffs and the Wetlands as a single diverse protected Reserve, an idea examined more closely in the final chapter.

There are other people who are thinking along related lines particularly with regard to marine reserves. The summer 2021 edition of Devon Wildlife Trust's *Wild Devon* refers to government plans to designate Highly Protected Marine Areas (HPMAs) by the end of 2022. *"If implemented successfully this will mean the arrival of the U.K.'s first ever sea sanctuaries where damaging activities will be banned"* Joan Edwards, head of "Living Seas" at the Wildlife Trusts told Wild Devon *"this special form of protection is vitally needed. Decades of over-exploitation and pollution have left our precious seas damaged and the wealth of wildlife that once lived there is much diminished. Existing MPAs are limited in their ability to restore nature, as they only go as far as conserving its current, sometimes damaged, state. HPMAs will allow us to see what truly recovering seas look like. They will set a new bar against which other protected areas could be measured".*

Following the government's announcement the Wildlife Trusts will be participating in the consultation process. They believe HPMAs should be designated in each regional sea whether English inshore, near-shore or offshore waters. This would include a range of habitats in which experts could study how ecosystems recover when pressures are reduced. Joan Edwards continued *"We know these highly protected areas will work. When bottom trawling was banned from Lyme Bay we saw the astonishing recovery of beautiful Sunset Corals and Pink Sea Fans within just a few years. When fishing was banned from Lundy, fisheries benefited from increased lobsters and tourism boomed following positive marine publicity".*

Triassic cliff east of Sidmouth.

Passing Cretaceous cliffs west of Beer.

Steps down through mudstones to the beach at Salcombe Mouth.

CHAPTER TWELVE
Achievements, reflections and the future

On the 11th December 2001 Marcus Binney, architecture correspondent of *The Times* wrote about World Heritage *"Four British sites are set to be added to the World Heritage List later this week. The four put forward by the British Government reflect a call by UNESCO for nominations of categories of sites which are not well represented on the list, which includes most of the acknowledged wonders of the world such as the Pyramids and the Taj Mahal as well as an increasing number of lesser-known but no less remarkable places.*

"Britain, as the cradle of the Industrial Revolution, has nominated New Lanark, the famous model industrial community developed by the pioneer Socialist Robert Owen who believed that well housed workers would produce better results; Saltaire, the model village built by the Yorkshire textile magnate Sir Titus Salt to house the workers in his mills near Bradford; and the historic textile area of the Derwent Valley which saw the origins of the factory system and mass production under the lead of the great Sir Richard Arkwright.

"The fourth site, the Dorset and East Devon Coast, reflects a call from UNESCO for more natural as opposed to man-made sites. This stretch of coast from Orcombe Point in Devon to Old Harry Rocks near Swanage reveals an almost unbroken series of exposures of sedimentary rocks laid down over 180 million years from the Triassic age to the Cretaceous, where fossils new to science continue to be found as the cliffs erode.

"21 members of the World Heritage Committee (which for the first time includes the UK) are likely to approve all 40+ nominations". Two days later the 95 miles of coast was declared a World Heritage Site as *"an outstanding example representing major stages of the Earth's history, including the record of life, significant ongoing*

geological processes in the development of landforms and significant geomorphologic and physiographic features."

A year before this Albert Knott had asked me to compile a dossier on Undercliff history and wildlife, and provided me with plenty of information from the English Nature files. Despite a thousand pages from Norman Barns and plenty of help from the Philpot Museum I also needed to interview those with specialist knowledge of the wildlife and/or management of the reserve. In July 2001 therefore, I visited the headquarters of English Nature in Exeter to talk with Phil Page, the Undercliff's site manager and then Albert who had written the third management plan in 1998 soon after moving to Devon to join English Nature. Although we would later work happily together and become friends I hardly knew him at the time. He had joined Phil at Yarner Wood in 1997 and having chalk grassland experience was invited to write the new Management Plan. He would be the first to admit that he found it difficult to express himself clearly on paper but thought that a plan produced by one person could be more effective than a team production like the previous one. Like other interviews carried out around that time it was tape-recorded so that quotations would be accurate.

"I was lucky to write the new Management Plan as the old one was written by four different people and I had to bring it up to the present day and give it a new structure in accord with present practice. Although it's repetitive I like the thought process. In writing the plan I was very aware of lack of knowledge; you can carry out feasibility studies on Holm Oak control or herbivore management but you have to know what's where; hence the reference to simple repeatable survey techniques. Ten years or five down the road you're going to need to repeat things ... So you've got to have visual checks or something that can be backed up with mini-surveys so that you can say that something has increased, gone, or is doing very well. My plan was based on the previous one with the reserve primarily managed by natural processes. The limited intervention approach isn't about doing nothing but involves looking ahead and allowing the natural processes to continue unhindered.

When I asked about the landowners and their level of cooperation he answered by reference to Judith Fiske who was genuinely interested in the reserve and questioned him on plenty of details at their annual meeting. She was clearly interested in insurance problems. *"When EN had the lease from Allhallows College we could have made more of* their interest. As soon as it turned into the Rousdon Estate the developer dashed in and could dash out again once money had been made. We will have problems with lots of new owner occupiers who need to be trained to respect the importance of the neighbourhood. Then we've got the Allhusens who are very approachable but have just contracted out the land management of the hinterland to the Bindon Estate with whom we will be discussing the possibility of a buffer strip. To be honest I think their agent is quick off the mark and sees that things are changing. I will continue to talk to him in the hope that he will agree to the buffer or come back at a later date especially as subsidies are going to be down'.

"Another landowner who can cause interesting challenges in the future is South West Water. They've bailed out some of their equipment but left the pumping station there. The water is no longer being managed through the set of pipes but flows into the toe and all the signs are that the whole area will go. I've got Peter Grainger to survey the area now. He will be able to report back to us on the state of movement since the winter of 2000/2001. From this we hope to have an accurate picture of what happens; a far better picture of where and why."

After some queries and answers about the chasm Albert moved on to Pinhay "If it seriously slips that will be the area to make vegetation surveys. On Friday, for example, I'll be going out to check on the soft rock pioneer features, looking at species change. We don't have much information about the soft rock pioneer communities; we will use quadrats for example on Pinhay Warren. Places that have either got fixed or disturbed soft rock communities are in a sense my prime habitat, along with the developing vegetated sea cliff scrub and Ash/Maple woodland. With this woodland I'll have to go back to the Management Plan and describe what it actually is; it's not pure Ash/Maple, it's blocks. Things like controlling Holm Oak aren't as important as those two."

I ask about the factors that will help him to select potential glades along the coast path. "My priorities are as follows: one is people walking through the site and the second is management of exotics; convenience will come third. Glades as you walk along are a public requirement, if their creation can be linked with management of Holm Oak, fine... As I mentioned to you, coppicing Hazel, not because of its importance in terms of the site, but with light getting in it's a contrast for people; that's the way things are going. We'll be trialling stock grazing if we get the landowners' cooperation, the *buffer strip is a real priority and the removal of the old pumping station would be a coup. If it doesn't go this year we'll have to pick up the pieces with South West Water. There'll be a lot happening in the next five years. There is so much work that could be done but we have to be pragmatic. You can chuck money at a site and create habitats but I have to follow this line of limited intervention. It's no good going in to "bust a gut" creating something that could be created by landslip tomorrow."*

When I suggest that more open habitat was created last winter than by any previous management Albert finished *"I'd rather it happened through landslips. I'll work on access and improving people's experience of the site. To that degree and thinking about interpretation, saying that paths are slipping, that there is no access to the beach and telling them how much further it is to Lyme, as well as site-related interpretation. Some of the information you have collected can be quoted. We'll have something, pictures, maps, background information to work on."*

At that time I gave five reasons for it to be the time to increase investment of time and money in the reserve. A prime reason was the inadequacy of past funding but the declaration of the coast as a World Heritage Site and Special Area of Conservation (SAC) both raised the status of the Reserve which was coming up to its 50th year at a time when it was desirable to have more public access and publicity about National Nature Reserves.

I also suggested 15 studies that might be useful starting with more research into landslips (1) and checks on the status of rare or local plants (2). Of these the Purple Gromwell and Yellow Bird's-Nest have not been recorded for twenty years nor Early Gentian for ten. Yellow-Horned Poppy is locally threatened by rising sea levels but Nottingham Catchfly has received recent help. Birds and butterflies have continued to receive attention but not in the form of regular, repeatable surveys (3) while "interesting" species are recorded by some with accurate map references (6) and by others, like me, with vague references to Dowlands or Haven Cliff. Bats, moths (10) and many invertebrate groups have received much more attention, at least at times, while lichens, Bryophytes and fungi (7) have all been subject to expert study but again with no consistent recording. I know of no detailed studies of the saproxylic invertebrates and their colonisation of dead wood (11) nor of succession following landslips (5) although one attempt at such was ruined by further land movements. I have no idea why I suggested that Haven Cliff top should be the place to survey invertebrates (9).

Achievements, reflections and the future

Nottingham Catchfly has received recent help.

There have been no regular, repeatable surveys of butterflies. What is the exact status of the Large Skipper?

The sheep-wash has recently been well maintained.

Vague references to Haven Cliff which extends over a mile.

227

Chapter 12

Recording in aquatic habitats (12) has been limited to flies, the Bioblitz and an occasional search for Great Crested Newts while coastal bird movements (13) are now concentrated on the more suitable location of Beer Head. Human artefacts (14) like walls, cottages and plantations receive desultory attention while the sheep-wash has been well maintained and its role and history highlighted. The public I talk to seem happy enough about most aspects of the Undercliff but whether they are aware of the role of Natural England (8) I know not; they can learn on our guided walks. Personally, I leave the insurance/liability situation to Natural England (14) while the final suggestion, of extending the NNR into Lyme Bay (15) was discussed in the previous chapter. It remains a priority.

These potentially useful studies have obvious links to the management policies but little glade creation (1) has occurred along the coast path. Of necessity a buffer strip along the cliff top (2) followed the rerouting of the coast path and the Allhusen's permissive path through Lynch Meadow and along to Pinhay has been a great gain. Access to the Chasm, Humble Glades and Haven Cliff (3) has been enormously improved. Display boards at several sites give walkers useful information (4). As described in this book much mowing and slashing (5) has occurred but I don't think spot treatment of Holm Oak seedlings before mowing has been tried. Pumping station buildings and pipes have been removed (7) some felling of Holm Oak occurred (8), Japanese Knotweed has had a hard time (10) and the future of the Wood White is receiving serious attention (9). Bracken is effectively controlled on Goat Island east (12) and the "screes" on the northern and eastern slopes of Goat Island are actively managed (11). Long circular routes for the tough walker are possible at both ends of the reserve (13) but the regular attempts to prevent bramble thickets around Humble Pond are not completely successful (14). As before the final suggestion was about the possible extension of the NNR area out to sea (15).

A few days before my meetings with Albert and Phil and my subsequent thoughts on management and recording, I had met David Allen while being shown some of the attractions of the Stockland turbaries. I knew little of his long association with the Undercliff other than his guardianship of Tom Wallace's extensive records. Twenty years later we have shared plenty of Undercliff exploration but I still wanted his version of recent experiences and unusual species found. With Covid-19 preventing a meeting I wrote asking a few questions.

Answering one about his favourite activities, he described the peak of his enjoyment is 'almost getting lost' when exploring new routes whether in the Chasm, going up to Goat Island by one of the two ways up from the old coast path, visiting remote Chapel Rock or climbing from the beach up Haven Cliff by the Elephants Graveyard and on to the cliff top. Just as good were the shared experiences of guided walks, particularly with Colin Dawes, and good field study days counting thousands of Autumn Gentians or searching for Cliff Tiger Beetles on Pinhay Warren. Another favourite was the satisfaction gained from taking part in decisions about the need for new clearance before conservation work parties, taking part in the work and seeing the results of team activity.

Cliff Tiger Beetle

Few of David's records were completely new but the Purple Hairstreak had been a first as had the discovery of the arable weed Weasel Snout 170 years after there had been farmers' fields on Goat Island. Leopard Marsh Orchid at Culverhole in 1962 and Marsh Fragrant Orchid there four years earlier were new records for the reserve as was Henbane found on Haven Cliff much more recently. Yellow Bird's-Nest in 1967 was probably the second Undercliff record.

"It is great to have the new and improving area of alkaline grassland in Humble Glades both for its own sake and as a link to Charton Bay and the unstable cliff habitat there. The glades could be difficult to maintain so careful thought is required before clearing much more. Further work on Goat Island is justified as the terrain is so much less challenging but maintenance may depend on volunteers unless government starts to fund Natural England in a sustainable way."

Autumn Gentian

Further afield David's favourite habitats are the bogs of the Blackdown Hills AONB and his most useful activity the 25 years of influencing their management, even if he makes no claims to be a committee man. He has also taken great pleasure from returning to family roots and being a local despite his years away overseas.

Rob Beard is another enthusiast but one whose experience of the area is comparatively recent. Access to the archives has really impressed on him the level of vegetation change in the years since the NNR's nomination. *"Old photographs, accounts and surveys provide clear evidence of the development of woodland over significant areas of calcareous scrub and grassland at places like Haven Cliff, Dowlands and Humble Green. Tom has really stepped away from the non-intervention policy and valuable areas of habitat have been restored or created but activity is always constrained by the limited workforce."*

Some of those areas are illustrated opposite. Clearance around the western end of Goat Island began on my 80th birthday, and in the four years since then extensive glades with plenty of edge effect have been opened up with scrub adjacent to some form of grassland. By 2021 the effect can be seen as Fuggles and I head towards the patch of recently cleared land described earlier. It is on the western extremity where it is possible to look over the precipitous mouth of the Chasm. The trees ensure that it is not a particularly impressive view.

Further east another open area is bordered by a less steep descent into the Chasm. As rain quickly percolates through the chalk it is a dry area but more than twenty years ago one of our dogs found that a hole, slightly up an old hedgerow tree, is often full of water. Fuggles is the fifth of them to have learned about this useful stopping point.

Rob feels that sad though it is, one consequence of Ash dieback will be that it may afford the opportunity to bring some areas back into open habitat with appropriate management. He also has clear views on Holm Oak *"which he would like to control in a decidedly forceful way, whereas Tom would argue that it has some landscape value which will increase with Ash dieback."* Perhaps this hesitancy is linked to public perceptions of tree felling.

Rob has been very active on Haven Cliff, an area frequently neglected in the past because of the difficult access. The new steps up from the Harbour have made a great difference but have also led to problems with litter and fires. The SSSI unit of Haven is still deemed to be in unfavourable condition but workers there see year-on-year improvements and will continue to seek additional funding. The cliff is a key site for the Wood White and the aim is to use the varied topography to create a mosaic of grassland, scrub, rock and woodland

Achievements, reflections and the future

Glade with plenty of "edge".

Not a particularly impressive view.

A less steep descent into the Chasm.

Fuggles after a drink.

with varying degrees of aspect, light and shade. Moribund anthills under Buddleia thickets give a strong clue as to the areas suitable for the re-creation of grassland. One of the most interesting aspects of the work has been clearing the low cliffs and limestone ledges of Privet, Ivy and the persistent Buddleia. Some of this has had to be carried out in a harness at the end of a rope. The cliff faces are great habitat for Nottingham Catchfly which is increasing while the Wood Whites survive on the, as yet, meagre patches of suitable habitat while also enjoying the shelter along the narrow path.

In Humble Glades observation suggests that the butterfly has a preference for the first two eastern glades, the first of which is underlain by Foxmould sand and has wet rivulets. Paul Butter emphasises the Wood Whites' liking of humidity so it might make sense to extend the area of open habitat to incorporate some boggy areas and small streams. At Humble Pond it would be good to link some of the smaller ponds, inaccessible for the last 15 years, to the main pond and to each other.

New access to Pinhay Warren from the boardwalk below Pinhay House could open up a completely new work front. This has involved making steep steps enabling volunteers to be taken down safely; AVDCS work parties were planned for the winter. Having found enormous populations of Bee Orchid and Marsh Helleborine Rob sees the Warren as potentially one of the most interesting areas of the reserve with plenty to explore and much self-sustaining exposed grassland.

Finally we come to Tom Sunderland who has had most influence over the reserve for 15 years although his activities have also extended far into Dorset. An interview was impossible early in 2021 but he sent detailed answers to my queries and was excited at the possibility of a book on the history of the Undercliff. Asked about his ideas for management action when he was first appointed he replied that he was *"overawed by the scale and remote wildness of the site and therefore wanted to get to know more about the area and about the characters associated with it."* He explained how he *"wanted where feasible to improve the public access to the reserve but to do this without compromising the wildlife interest or by creating too much disturbance or spoiling the intimacy and quietness."* He also knew that the burden of non-native species was far too high and that he wanted to extend the area of species-rich chalk grassland.

Asked about his favourite features of the site and his role in maintaining them he commented on *"the grand scale of the reserve and its associated remoteness and inaccessibility. In particular being able to visit, manage and introduce other people to these remote, out of the way and largely hidden "special places" such as Humble Glades, the Plateau and the Chasm is a joy and privilege. I have always seen my role on the Undercliff (and also in looking after the other NNRs that I'm responsible for) as something of a caretaker role – you want to hand them on to the next incumbent better than you found them ideally, and certainly not poorer or in a worse state. I have always particularly enjoyed taking groups of unsuspecting members of the public on guided walks with you and David."*

Tom has always feared that Ash dieback would affect the reserve very significantly. This has yet to be seen although the signs are not good as many of the trees are quite seriously infected and felling has begun. He has spoken with Emma Goldberg, Natural England's woodland specialist and also, in an informal capacity, with Keith Kirby after he left Natural England. *"The general consensus is that a non-intervention approach is the most appropriate management for the Undercliffs. There are reports of Ash rejuvenating and recovering and this may happen depending on local genetic resilience or susceptibility. If the resistant strain was identified either from the NNR or elsewhere then it might be feasible to carry out some sort of recovery "cluster planting" whereby resistant trees were planted in small clusters across parts of the reserve with the aim that they would spread naturally through the reserve over the years. This would only be done as a last resort."*

Changing the subject, I ask whether he is happy with the changes along Haven Cliff and whether the coast path will run through that way. *"The work that Rob Beard has carried out over the last couple of years, built on previous work by contractors, is fantastic and is restoring the area wonderfully for wildlife. The plan is to continue removing Buddleia and other scrub to create a scrub/grassland mosaic."* Tom is hopeful that Natural England may be able to introduce some form of grazing animal, possibly a small herd of goats, which could be constrained by satellite grazing collars (therefore no fences would be required) with water supply run from the golf course above.

Answering the second part of my question he thinks it unlikely that the coast path will go through this part of the site. *"It has been carefully considered as part of the coastal access provisions but was ruled out on health and safety grounds by the access team because of the active landslip/coastal rockfall approximately 150 m west of Finger* and Thumb. If that stabilises it might be possible to reconsider the route but the "squeeze" there (by which he means the short distance between the foreshore and the inland cliff) is very tight and the gradients are consequently steep with few options for path routes. The inherent instability associated with the steep gradient is an issue so it may have to remain an unofficial permissive access route braved by the few."

In answer to another question about "new experts" Tom tells me that he is currently *"working with the University of Plymouth and researchers who are developing an early warning system for landslides. This involves drilling into boulders and installing GPS locators that can pick up and measure even the slightest movement of the boulder. I would also like to see a more detailed map developed of the landslip complexes across the reserve and how (and if) they interact, and, if it were possible, to develop a way of making accurate predictions about the more significant landslides and where, and when, they might be expected to happen."*

An assemblage of lichens.

He would like to have helped more with the listing of lichens and bryophytes but all species need to be recorded. Improving habitat for the Wood White and monitoring the species across the reserve is important as the Undercliff may be, or may become, a national stronghold for the species.

There is a simple answer to my next question about the frequency of management plans. They are required by Natural England for all NNRs so Tom has little choice but does agree that every five years is more than enough. He sees possible difficulties with the resources needed for any extension to Humble Glades for some parts that had been cleared have quickly reverted to bracken, nettles and scrub. However *"given resources anything is possible and I would not be averse to widening the corridor"*. Resources would be equally necessary if Pinhay Warren were to be made accessible again so that the rapidly spreading gorse could be controlled and the rich array of invertebrates investigated again. Tom, together with Paul Butter and Doug Rudge had visited recently partly to find or create a good way in and partly to plan the future work. In the longer term, after gorse clearance, a walk through the Warren to Underhill Farm could eventually be established.

Finally, regardless of any questions, Tom wanted to mention an initiative that fits in well with my own thinking. *"There are developing ideas around an East Devon Undercliffs Nature Recovery Network which would involve a number of local partners including the National Trust, East Devon District Council, East Devon AONB, the Jurassic Coast Trust and Devon Wildlife Trust. The network would seek to improve management of the coastal undercliffs from Lyme to Sidmouth, manage and extend areas of coastal grassland, control non-native species, join up existing wildlife sites and aim to secure long-term de-intensification of arable land inland via targeted agri-environmental options. There would also be a focus on the importance of the geology and its interpretation, increased "citizen science" and volunteering to support monitoring of species and habitats and the possibility of extending the National Nature Reserve status to Branscombe Underhooken and beyond."*

These thoughts about the future promote thoughts about the past, and those pioneers who saw problems which would arise. Water levels and water quality present intractable problems all over the world today, but this was already appreciated by explorer Alexander von Humboldt in 1800. In *The Invention of Nature* Andrea Wulf comments on his views on the falling water levels in Venezuelan lakes. He believed that the drying out of Lake Valencia and the consequent problems for local inhabitants was due to clearance of forests as well as the diversion of water for irrigation. Having lost the forests *"heavy rain is no longer able to sustain the trees"* but instead *"bore down the loosened soil and formed those sudden inundations that devastate the country."*

The author went on to comment how frequently Humboldt observed that *"humankind unsettled the balance of nature"*. Deep in the Orinoco rainforest Spanish monks lit their ramshackle churches with oil gathered from turtle eggs but as they never left any eggs to hatch the turtle populations collapsed. Earlier Humboldt had noted how uncontrolled pearl fishing had depleted oyster stocks. It was *"all an ecological chain"* and *"everything is interaction and reciprocal"*.

On his extensive travels he frequently witnessed vicious contests between predator and prey. This was the web of life *"in a relentless and bloody battle"* a concept very different from the prevailing view of nature as a well oiled machine in which every animal and plant had its divinely ordered place in the service of man.

Although his observations were made far from Europe's civilised court life to which Humboldt returned, his books continued to introduce controversial ideas. When he eventually fulfilled an ambition of travelling across Russia he seemed happy to break many of the limitations imposed upon his movements and to pay little attention to the agents of the Czar. In less than six months of extended exploration he and his party travelled 10,000 miles. Later his books about the expedition deplored the destruction of forests, the ruthless irrigation and the great volumes of steam and gas produced by the new industries. In a speech on his return to St Petersburg he encouraged the climate studies which would soon take place across the vast Russian empire.

Later the book describes how Humboldt played a vital role in the life of Ernst Haeckel who read Humboldt's books as a boy. He was desperate to read *Cosmos* as soon as he got home from school. Later, on his 25th birthday he heard of Humboldt's death and was miserable. At that stage Haeckel hoped to study marine life in the Bay of Naples and to make discoveries among the sea urchins and starfish there. Months later he found the project he needed, a study of radiolarians in Sicily. After reading *The Origin of Species* he became an ardent promoter of Darwin's views. Following his wife's sudden death in 1864 he came to live like a hermit, to become obsessed with evolutionary ideas and to work tirelessly on his two-volume *General Morphology of Organisms* which was a rallying cry for evolutionary studies in Germany. In the book he was the first to use the term 'oecologie' based on the Greek 'oikos' - a household. The word was then applied to Humboldt's study of the natural world. Apparently Haeckel was overawed when he visited Darwin in 1866 but this didn't stop him from spreading the idea of natural selection most effectively in his books written when he was back in Germany. He believed that nature could only be seen as one unified whole, a completely integrated kingdom of life. Later, in Java, he observed the intensity of the *"struggle for existence"* as plants and animals lived in vicious competition with each other and with their symbionts and parasites. It was Humboldt's *"web of life"*.

Some great admirers of Humboldt were based in North America. They included George Perkins Marsh who was born in 1801, Henry David Thoreau and John Muir. Late in life, in 1864, Marsh had written *Man in Nature* expressing the views that *"all nature is linked together by invisible bonds"* and that *"humankind is destroying the earth"*.

Thoreau, born in 1817, wrote seven very different versions of *Walden* between 1847 and 1854. The book which later became very popular, called for the preservation of forests and bemoaned the unrestrained misuse of water and the *"ruthless greed as irrigation diminished great rivers and made soil saline and infertile"*.

John Muir (1838 - 1914), born in Dunbar, was always ready to work tirelessly for the places he loved. His father, believing that the Church of Scotland was tainted with too much orthodoxy, wanted to be his own priest and to find religious freedom in America. In Scotland he never allowed pictures, ornaments or musical instruments in the home but his wife was able to find beauty in the garden and their children roamed the countryside. When not himself exploring, Muir had read about Humboldt's explorations and *"dreamed himself to exotic places"*. Once in America he continued to wander, settled for a time at the University of Wisconsin and, avoiding conscription, escaped to explore Canada when the American Civil War began. When it ended he returned to America, walked through five states and grew closer to Humboldt in spirit while experiencing the natural world in a new way.

As America changed over the next 25 years, in each of which 15 million acres were claimed for agriculture, while railways and industrialisation spread with people moving to the new cities as well as to the farms. Muir returned to Yosemite time and again. Each time he was shocked by the changes, for although it was a state park, control was lax. He turned to activism and campaigned for a national park comparable with Yellowstone in Wyoming, the first one ever to be established. Eventually in October 1890 nearly 2 million acres were set aside by US Federal, not Californian control, but the valley remained Californian. By the turn of the century Muir had gained so much influence that even the President Theodore Roosevelt wanted his company and camped with him in Yosemite. Muir managed to convince him that the valley should join the rest of the national park under the control of the Federal

Muir and Theodore Roosevelt at Glacier Point, Yosemite (1903).

Government. Roosevelt kept his promise and in 1906 the valley became part of the national park.

80 years later Muir's achievements were remembered in his native Scotland when a Trust in his name was set up to acquire and manage wild land. More recently land in the Lake District has come under Trust management but its membership extends much further south, with an active group based in Bristol. In 2019 I was invited to give a talk about the Undercliff at the group's AGM. In Scotland my grandchildren frequently join activities on Trust land in Knoydart not far up the coast from their home in Morar. In the South, East Devon Countryside staff lead John Muir events as part of a varied programme held in Local Nature Reserves. In *My First Summer in the Sierra* Muir had written *"when we try to pick out anything by itself we find it linked to everything else in the universe"*, another version of George Marsh's assertion that *"all nature is linked by invisible bonds"*.

National Parks, like John Muir Trust properties, stop at the coast but an article in the magazine of the Campaign for National Parks asks whether this has to be. In *Points of View*, Natasha Bradshaw comments that eight of Britain's National Parks have some coastline but that none of them include any sea. She believes that *"people's appreciation of the sea has been left behind and that maritime cultural heritage needs stronger recognition. The Blue Marine Foundation, "BLUE", proposes a commercially led process identifying (potential) National Marine Parks. They would encourage the creation and awareness of opportunities for people visiting the seaside, together with management of our coast and seas."* BLUE's vision is based on the experience of the Plymouth National Marine Park which was self-declared by 70 stakeholders in 2019 based on the opportunities it would bring to the people of Plymouth and to local marine habitats. *"National Marine Parks have the potential to take us a new distance towards engaging people with the seascape in a similar way that landscapes are enjoyed and protected for the benefit of people through National Parks"*.

In addition to the potential for Marine Parks a landscape review in 2019 suggested three new National Parks in England; the Chilterns, Cotswolds and the county of Dorset were the potential candidates. In another sign of change in 2020, the government announced that it planned to create more National Parks and AONBs to help meet its commitment to protect 30% of the U.K.'s land by 2030. Sandra Brown suggested that a National Park in Dorset would *"not only embrace the spectacular UNESCO World Heritage Coast but also double the*

Achievements, reflections and the future

Looking east at Stonebarrow

Both Stonebarrow, on the Golden Cap Estate, and Durdle Dor would be part of a Dorset National Park. If it extended into Devon the Park would include more spectacular coastal locations such as the Triassic stacks at Ladram Bay, the Cretaceous cliffs west of Beer ...

... and fossil-rich Jurassic rocks below the Undercliff.

currently underrepresented coastline in English National Parks". Other people, including Michael Dower, son of John Dower, advocate for the Parks in 1945, support the proposal of the Dorset and East Devon version *"which could fulfil the earlier vision and benefit the environment, the economy and communities of this very special area"*.

In the context of the proposed East Devon Recovery Network and less intensive management of arable land, it has long been clear to some that agricultural chemicals have effects far from their point of application. Apart from taking precautions in their delivery, buffer zones where no chemicals are used would have extensive environmental and recreational benefits

Much of the coast path through the Undercliff is relatively chemical free but sadly not the Chasm. Some 80 years ago Norman Barns challenged the then widely held view that the 1839 landslip had deposited "virgin" soil, whatever that might be, into the Chasm. In fact the land all round would have been treated with lime from a nearby pit, now marked by the trees growing in it, and by the dung of farm animals. When Norman first walked over these fields the animals were sheep and pigs. At that time, in the 1930s, woodsmen below were making hurdles and cutting posts and stakes for use on the estate above. The Warren below Pinhay, which had had a warrener's cottage, was no longer providing rabbit meat on the scale that it had done but below Whitlands the two fields, the pheasantry and orchard were still in use in the grounds of West Cliff Cottage. When the war came the large fields at Donkey's Green were dug up and cultivated intensively as part of the "dig for victory" campaign. Parts of the Plateau were used as allotments. No, said Norman, the so-called virgin naturalness was just an abandoned tract of farmland and countryside when it was proposed as a National Nature Reserve. Since then the run-off and percolation of soluble agricultural chemicals has made some of the reserve so fertile that nettles up to nine feet high dominate parts of the Chasm.

It is not only fertilisers that affect ecosystems, as chemicals to control insects, fungi and weeds have long been used. As early as 1763 nicotine was tried as an insecticide and by the 1800s grape growers on the continent were using compounds of sulphur and copper in attempts to control mildews. Paris Green, containing copper arsenite, was in early use against Colorado Beetle as was lead arsenate later in 1892. Organo-mercury fungicides were used in the early 1900s. DNOC, a synthetic organic herbicide introduced in the 1930s was poisonous to mammals

Achievements, reflections and the future

whether by inhalation or absorption through the skin. MCPA, an early hormone-based weedkiller, had properties that exaggerated the effects of natural growth-regulating substances leading to the treated plants outgrowing their strength.

Pests of all sorts had therefore been targets well before the 1940s brought in new compounds like BHC or ones whose insecticidal properties had only recently been discovered such as DDT. Their benefits were initially unquestioned but by 1960, when Norman Moore's role changed as he came to lead the Toxic Chemicals and Wildlife Section of the Conservancy, there were growing questions to be answered. He was soon able to describe some of the hazards faced by wildlife explaining how the worldwide use of DDT against Codling Moth and other orchard pests killed the predators that had previously kept harmful mites in check. Without their predators Red Spider Mites became major pests around the world. The Toxic Chemical Section also had the problem of finding out which pesticides were particularly hazardous to wildlife. Well before Rachel Carson's *Silent Spring* had

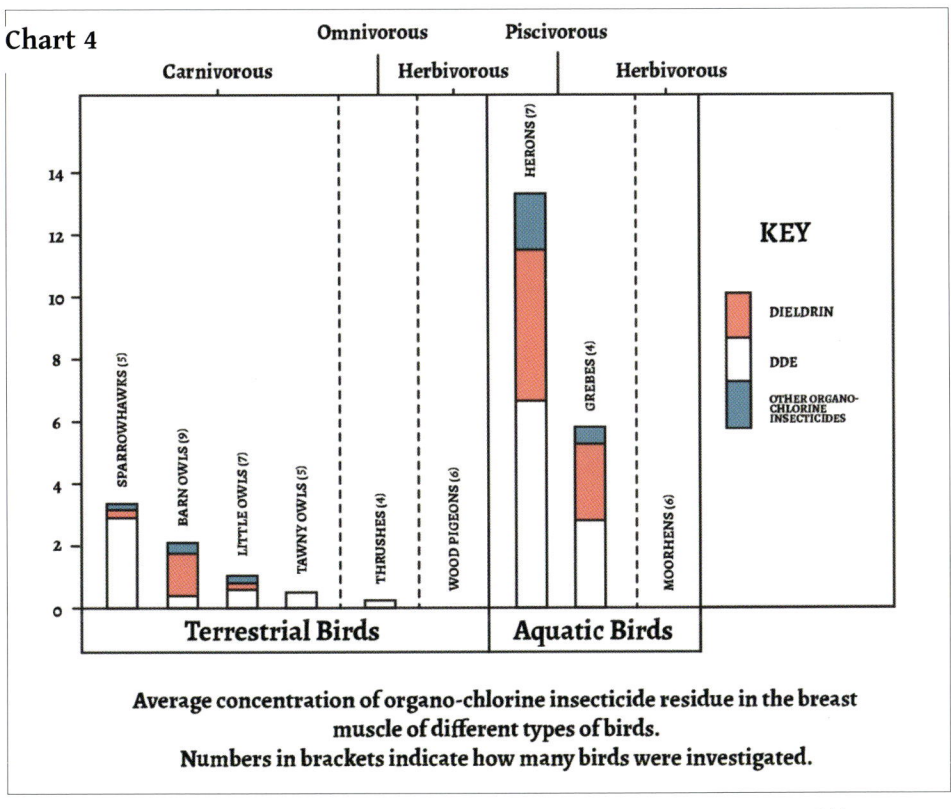

Chart 4

Average concentration of organo-chlorine insecticide residue in the breast muscle of different types of birds.
Numbers in brackets indicate how many birds were investigated.

made Clear Lake and its Western Grebes famous, Charles Elton had drawn attention to a little-known Californian journal which had reported how a mildly toxic insecticide had been used to control an irritating gnat with disastrous consequences for the grebes. Analysis of the concentration of pesticides at different levels of the food web convinced Norman Moore that their persistence could be as important as their toxicity and that species at or near the top of the food web were most vulnerable.

Two things were needed, dead animals from around Britain and a method of analysing their bodies to find the concentration of pesticides. The first attempts used paper chromatography but the development of gas/liquid chromatography made analysis quicker and more accurate. Many of the analyses were carried out by Colin Walker, much later a committee member of the local conservation society. A paper in *Nature* in March 1964 included Chart 4 on page 241, which shows the concentrations of organo-chlorine insecticide residues in the breast muscle of different bird species. The muscles have much lower concentrations of insecticides than the fatty tissues had.

It was soon shown that other factors, such as the permeable gills of fish could be critical. It also emerged that a single relatively obscure species could actually be vital in maintaining a local ecosystem. This was the case with the millipede, *Glomeris*, breaking down the dead leaves of

Ice melts and sea levels rise - a calving iceberg in Antarctica

invasive Tor Grass before microorganisms were able to reduce the leaves still further making nutrients available throughout the system. Later studies show that in the absence of *Glomeris* other millipedes moved in.

Tor grass is an aggressive and unpopular coloniser of Undercliff grassland, suppressing many of the small calcicole plants that help to make chalk grassland so attractive. Recently there has been a reassessment of the management regime with the suggestion that the practice of cutting all the grass in September removes habitats for potentially overwintering invertebrates.

Locally and linked to the earlier reference to the Nature Recovery Network, it would seem that the inshore waters of parts of Lyme Bay should be included in the ambitious conservation scheme.

The complexity of ecosystems and the interaction between their components were being shown to be more complex than the American pioneers had suggested, and the idea of protecting nature reserves or other habitats through a non-intervention policy was seen to be a failure, as natural succession would inevitably lead towards scrub and then tree domination in most British conditions. Now in post Common Market Britain agriculture is set for big changes. With so many farms being relatively small, with many farmers growing old and with the subsidies that made their lives possible going or gone, it could well be time for radical changes in effective land use. Knowledge of natural processes is still limited but we do know that we must protect and retain a wide range of species not only for their own sake but also for their part in the maintenance of a stable environment. Sadly and disastrously the world's environments are far from stable as they grow warmer, as ice melts and sea levels rise: the changes will have to come quickly.

BIBLIOGRAPHY

Allen, David J (2000) *Wildflowers of the East Devon Coast.*

Allen, David and Lock, Mike (2016) *Introduction to the Undercliffs in 'Flora of Devon.'* "Botany of the Axmouth to Lyme Regis Undercliffs".

Allhusen, George and family (2020) Rainfall records at Pinhay 1868-2020.

Ambios Environmental Consultants, (1995) *Lyme Bay Environmental Study.*

Anon (1840) *A Brief Account of the Earthquake.*

Anon (1840) *The Extraordinary Landslip near Axmouth* (in the *Penny Magazine* of the Society for the Diffusion of Useful Knowledge, 15th Feb 1840).

Anon (1840) *Poetical Remarks on Hearing of the Great Landslip.*

Anon (1840) *The late Melancholy Accident* (in the *Mercury*).

Arber Muriel (1939) In *Country Life* "The Great Landslip of 1839".

Arber Muriel (1940) *The Coastal Landslips of South-East Devon.* Proc. Geol. Ass. (51)

Arber Muriel (1973) *Landslips near Lyme Regis.* Proc. Geol. Ass. (84)

Arber Muriel (1988) *Lyme Landscape with Figures.*

Archibald, Hamish (1965) *The First Undercliffs NNR Management Plan.*

Armitage, Patrick (1969) *The Littoral Fauna below the NNR.* Proc. Dorset Nat. Hist. & Arch. Soc. Vol. 91.

Armitage, Patrick (1983) *The Invertebrates of some Freshwater Habitats in the NNR.* Proc. Dorset Nat. Hist. & Arch. Soc. Vol. 105.

Axmouth Study (1971) *Plans for the Development of Marshland around the Axe.*

Axe Vale and District Conservation Society, (from 1971) *Newsletters.*

Baring Gould, Sabine (1899) *Winefred, a story of the Chalk Cliffs.*

Baring Gould, Sabine (1890) *A Book of the West.*

Barnes, presumably William (Jan 1840) *Letter to his father* describing a visit to the Landslip.

Barnes, William (1842) *The Narrative of a Tour in the Splendid Summer of 1842.*

Barns, Norman (1980) A Day in the Undercliffs (*Nature Conservation Newsletter*).

Barns, Norman (1985) *Guidance Notes for Walk Leaders.*

Barrett, John and Yonge C.M, (1950) *Pocket Guide to the Seashore.*

Bath Journal (20/1/1840) *Experience of a Coast Guard as the Great Landslip starts.*

Bickley, F. (1911) *Where Dorset meets Devon.*

Bradshaw, Natasha (2021) National Parks – Why stop at the coast? (in *Viewpoint 2021*).

Bridport and Lyme Regis News (12/04/1929) *Report of improvement to the Coast Path.*

Bristow, Roger (1993) *Devon Butterflies.*

Brown, Roland (1857) *Beauties of Lyme Regis and Charmouth.*

Brown, Sandra (2021) *New National Parks* (in *Viewpoint 2021*).

Brunsden, D. and Pitts, J (1973) *The Coastal Landslides of Dorset.*

Butler, Jeremy (Ed 2000) Peter Orlando Hutchinson's *Travels in Victorian Devon.*

Campbell, Donald (1999) *The Encyclopedia of British Birds.*

Campbell, Donald (2006) *Exploring the Undercliffs.*

Campbell, Donald (2008) *A History of the Birds of the Axmouth to Lyme Regis Undercliffs NNR*, in *Devon Birds* vol 61 No. 2.

Campbell, Donald (2020) *Rocks and Wildlife Around the Axe.*

Carson, Rachel (1962) *Silent Spring.*

Conybeare, Rev William (1840) *The "remarkable convulsion" described in two papers.*

Cook, Phil (2014) Entomology Section of *Devonshire Association Transactions.*

Cooke, Michael and Gibbs, David (2003) *Plant survey of soft cliff habitats.*

Coxes J and W (1993) *Plant Surveys of Goat Island and the Plateau.*

Coxes J and W (1995) *Plant Survey of Culverhole.*

Cramp, Stanley (1977-87) Editor *Birds of the Western Palearctic.* 9 vols 1977-94.

Darwin, Charles (1859) *The Origin of Species.*

Dawes, Dr Colin (2003) *Fossil Hunting Around Lyme Regis.*

Dawes, Dr Colin (2005) *Bird Watching where Dorset meets Devon.*

Dawes, Dr Colin (2006) *Rock-pooling around Lyme Regis.*

De la Beche, H.T. (1830) *Sections and Views Illustrative of Geological Phenomena.*

Dennis, R. W. G., Orton, P. D. & Hora F. B. (1960) *The New Checklist of British Agarics and Boleti.*

Devon and Dorset County Councils (and Prof Denys Brunsden 2000) *Nomination of the Dorset and East Devon coast for inclusion on the World Heritage List.*

Edmonds, Richard (2020) *Goat Island and the Chasm explained.*

Edwards, Rev. Z. I. (1862) *The Ferns of the Axe and its Tributaries*

English Nature (1995) *Invertebrate Site Register.*

Ford, E.B. (1945) *Butterflies. New Naturalist* number 1 (NN1).

Fowles, John (1969) *The French Lieutenant's Woman.*

Fowles John (1989) *Wormholes.*

Franks, Elaine (1989) *Sketchbook of the Undercliff.*

Freeman, Clare (1989) *An Ecological evaluation of the Undercliff.*

Gallois, Ramues (2006-2009) *Axmouth to Lyme Regis Undercliffs* (in three parts).

Gilbert, Oliver (2000) *Lichens* NN86.

Grainger Peter et al (1985 and '95) *Cliff movement around the Undercliff Pumping Station.*

Gray, Todd (1998 Ed.) *Travels in Georgian Devon.*

Gray, Todd (2000 Ed.) *Traveller's Tales, East Devon.*

Griffiths, I. (1967) *Inside England on the borders of Dorset and Devon.*

Haeckel, Ernst (1866) *General Morphology of Organisms.*

Harvey, Prof LA and St Leger-Gordon, D (1953) *Dartmoor* NN27.

Hawksworth, David et al (1995) *Dictionary of Fungi.*

Hayward, Peter (2004) *Seashore* NN94.

Hearne, Thomas (1710-12) *The Itinerary of John Leland, the Antiquary.*

Hebditch, Max (In *Museum Friend Issue 34, 2019*) – The Reverend Edward Peek.

Humboldt, Alexander von (1845-59) Five volumes of *Cosmos.*

Hutchinson, Peter Orlando (In *Saturday Magazine 8/2/1840*) *A Guide to the Landslip.*

Jukes Brown A J (1900) *Undercliff Cliff profile.*

Kilvert, Francis (1871?) *Diary*, edited by W. Homer, 1938-40 in three parts.

Knott, Albert (1998) *3rd Undercliffs Management Plan 1998-2003.*

Knott, Albert (2003) *4th Undercliffs Management Plan 2003-2008.*

Knott, Albert (2005) *Letting Things Slip in the Undercliff.*

Marren, Peter (1999) *Britain's Rare Flowers.*

Marren, Peter (2nd edition 2005) *The New Naturalists* NN82.

Marsh, George Perkins (1864) *Man on Nature.*

Mc Gowan Chris (2002) *The Dragon Seekers.*

Mellanby, K (1967) *Pesticides and Pollution* NN50.

Moore, Norman (1987) *The Bird of Time.*

Morris, FO (1877) *History of British Butterflies.*

Muir, John (1911) *My First Summer in the Sierra.*

Mumford (1898) *Illustrated Seaton, Beer and Neighbourhood.*

Murray (1859) *Handbook for Travellers in Devon and Cornwall.*

Naylor, Paul (3rd Edition 2012) *Great British Marine Animals.*

Nicholson, E.M. (1936?) *A National Plan for Britain.*

Nicholson, E.M. (1981) *Birds and Man* NN17.

Pitts John (1974) *A Survey of Historical Documents relating to the Bindon Landslip.*

Pitts John (1974) *The Bindon Landslip of 1839.*

Pitts John (1981) *The Landslides of the Axmouth-Lyme Regis Undercliffs.* PhD thesis, Imperial College, London.

Pogson, Margaret (2015) *My Devonshire Year.*

Proctor, Michael, Yeo, Peter (1978) *The Pollination of Flowers.* NN54.

Proctor, Michael, Yeo, Peter and Lack, Andrew (1996) *The Natural History of Pollination.* NN83.

Proctor, Michael, (2013) *Vegetation of Britain and Ireland.* NN122.

Roberts, George (1835?) *History of Lyme Regis.*

Rogers, David et al (1992) *2nd Undercliffs Management Plan.*

Sheehan, E.V. et al (2013) *Recovery of a Temperate reef Assemblage in an MPA.*

Snow, D.W. and Perrins, C.M. (1998) *The Birds of the Western Palearctic* (2 vols).

Spooner, Malcolm (1979) *Presidential Address* to the Devonshire Association.

Stamp, Sir Dudley (1969) *Nature Conservation in Britain.* NN49.

Stirling, D.M. (1838) *Guide to the Watering Places on the SE coast of Devon.*

Sunderland, Tom. (2010) *5th Undercliffs Management Plan, 2010-2015.*

Sutton, Stephen (1972) *Woodlice.*

Thomas, Jeremy and Lewington, Richard (1991) *Butterflies of Britain and Ireland.*

Thompson, Des, Birkes Hilary and John (2015) *Nature's Conscience* (Derek Ratcliffe).

Thoreau, Henry David (1854) *Walden*.

Tinbergen, Nikko (1953) *The Herring Gull's World* NN monograph.

Transactions of the Devonshire Association (various articles and dates from 1992).

Wallace, T. J. & Orton, P. D. (1993) *Some Agarics and Boletes in East Devon*. (Devonshire Association).

Ward, B.T. (1953) *Undercliffs Plant Survey* (before NNR declaration).

Watson, L.J. (30/9/49) *Notes on the Undercliff as a potential NNR*.

Western Morning News (2 and 26/8/39) Views on the purchase of part of the Undercliff.

White, Walter (1865) *England from the Thames to the Humber*.

Whitehouse, Andrew (2007) *Managing Soft Cliffs for Invertebrates*.

Wild Devon (Joan Edwards) (2021) *Good News for Marine Wildlife*.

Woodward, H.B. (1889) *Diagram of Section across Landslip*. (28/4/1889).

Wulf, Andrea (2015) *The Invention of Nature*.

Yonge, C.M. (1949) *The Seashore*. NN12.

Young, Geoffrey (1992) *Watching Wildlife*.

Index

Mainly people and localities, with some geological features, selected taxonomic groups and individual species

A

Acid soil 17, 35, 153, 169, 197

Aculeate wasps and bees 92, 125, 134

Agarics (fungi) 34, 108, 110

Allen, David ix, 39, 79, 108, 110, 118, 144-146, 150-151, 163, 165, 170, 177, 179, 228, 245

Allhallows school/college 17-21, 33-36, 38-39, 41, 56, 69, 99, 101, 115, 118, 224

Allhusen family vii, 12, 19-20, 30, 34, 72, 121, 156, 225

Ambios Environmental 113, 212

Ames, John 10, 13-15, 58, 196

Ammonites iv, 8, 72, 153-154, 203-204

Anabat detector 187

Anning, Mary 14, 91

Ants, anthills 93, 125-126, 131, 138, 143, 149, 152, 185, 198-199

AONB 25, 79, 166-167, 171, 196, 229, 235, 238

Arber, Muriel 9, 23, 30, 44, 56-57, 74, 76-77, 90

Archibald, J F (Hamish) 11, 35, 85-86, 96, 104, 106, 116, 121, 194

Armitage, P D (Patrick) 37, 119, 130, 211-212

Ash dieback 230, 233

Austen, Jane 4

Axe Vale & District Conservation Society vii-viii, 10, 21, 116-117, 146, 151, 203, 242

Axe, river 1-3, 6, 10, 19, 21-22, 74, 76, 102, 107, 116, 146, 160, 177, 203-204, 220

Axmouth 13, 1-2, 19, 30, 34, 51, 58, 74, 77, 87, 89, 130, 156, 160, 174, 207

Axmouth Study 87

B

Badman, Tim 169

Baker, Val 56, 112, 127-128

Bampfylde, Copplestone Warre 5

Baring Gould, Sabine 1, 55, 170

Barnes, Rev. Frederick & Son 63, 65

Barns, Norman vii, 10, 13, 19, 21, 40, 54, 56, 59, 74, 76-78, 89, 91, 93-94, 96-97, 99-101, 103-104, 106, 114-115, 117, 119, 121-122, 131, 133, 141, 147, 153, 170, 212, 224, 240

Bastow, Rev. R F 33

Bawden, Liz Anne 44

Beard, Rob iv, ix, 153, 159-160, 196, 215, 230, 233

Beche, H T De la 82

Beer 28, 54–55, 63, 89, 101, 153, 207–208, 217, 222, 228, 239

Bees 92, 125, 127–130, 138, 140, 142–143, 151–154, 173, 185

Beetles 10, 92–93, 120–121, 125, 130–132, 137–138, 140–143, 151, 171–172, 179, 192, 229, 240

Benfield, Barbara 40, 113–114, 118, 163–164

Benn, Jeff 40, 109, 164

Bindon 22, 28, 34, 44, 47, 49, 51–52, 54, 60, 64, 74, 76, 86–87, 92, 113, 133, 225

Bioblitz 108, 130–131, 163–165, 183–186, 189, 213–214, 217, 228

Biodiversity Action Plan (BAP) 137, 167, 189

Blue Marine Foundation 160, 238

Boletes (fungi) 108–109, 179

Bolton, David 119, 127, 131–132, 139

Bracken 9–10, 82, 86, 152, 186, 196, 228, 234

Bristow, Roger 165

Brown, Roland 10, 15

Brunsden, Prof. Denys ii, viii, 169

BTO (British Trust for Ornithology) vii, 26, 33, 35, 122, 133, 174

Buckland, William & Mary 29, 44, 46–47, 49, 51, 56–57, 59, 64

Buglife 137–139, 168

Butler, Nic 167

Butterflies & butterfly conservation 28, 32, 39–40, 93, 115, 139, 141, 144, 151, 156–158, 164–165, 174, 176–178, 183, 185, 192, 226–227, 232

Butters, Paul 156, 192, 232, 234

C

Chalk 1, 6–7, 13, 16, 19, 29, 33, 38, 44, 47, 62, 70–72, 81, 88, 93, 95, 97, 101–103, 116–117, 125, 141, 146, 150–151, 154–157, 170, 189, 203, 224, 230, 232, 243

Chapel Rock 109, 229

Charmouth 7, 71, 192

Charton Bay 31, 74, 77, 86–87, 106–107, 109, 113, 118, 124–125, 143, 145–146, 166, 169, 172, 174, 183, 187, 203, 208, 213, 215, 229

Charton Goyle 12

Chasm, The iii, 21, 28–30, 33, 43–44, 47, 49, 51, 54–56, 58–67, 86, 100, 107, 118, 133–134, 147–148, 163, 166, 170, 174, 197–198, 225, 228–231, 233, 240

Chert 72, 154, 203

Cnidarians 213, 215, 217

Coelenterates. See *Cnidarians*

Collins, Phil 169

Colyton 51, 63–64, 110, 169

Combpyne viii, 56–58, 88, 107, 115, 163, 165, 171

Conifers 11, 27, 35, 104, 186

Conservation iii, vii, 12, 25–28, 31, 33–34, 36–37, 56, 85, 87, 90, 92–94, 96–99, 102, 116–118, 122, 126–127, 137,

139, 143, 146-147, 149, 151, 156-157, 165, 168-171, 176, 182-183, 189, 208, 220-221, 226, 229, 241-243

Conybeare, Rev. William 14, 28, 43-44, 49, 51, 59-60, 63

Cooke, Michael 143, 145, 149, 151

Cooper, Toddy 125, 183, 185

Cox, W & J 108-109, 145, 149

Cretaceous iv, 16, 23, 70, 72, 161, 203, 222-223, 239

Crimea Seat 14-15

Critchard, Roger, Kath & Elizabeth 63, 88, 169

Culverhole 21-22, 31, 40, 70, 74, 76, 82, 108-109, 117, 124, 139, 143-145, 159-161, 172-174, 177, 179, 191, 207, 211-213, 229

Cumberland, George 16

Cuvier, Georges 14-15

D

Davis, Chris viii, 217-219

Dawes, Colin viii, 203, 229

Dawson, William 44, 47, 51, 59-61

Deer 36, 102, 106, 134, 167

Devon Fly Group 192

Devon Wildlife Trust 35, 129, 163-164, 220-221, 235

Devonshire Association 35, 118, 129, 165, 170, 189

Diptera (two-winged flies) 118, 125, 130, 134, 137, 165, 172, 185, 192, 228

Diver, Cyril 25-26, 28, 30

Donkey's Green 240

Dormouse 96, 160, 167, 186

Dorset viii, 5, 14, 29, 49, 53, 55, 95, 98, 114, 130, 137, 153, 173-174, 189-190, 204, 206, 211, 223, 232, 238-240

Dower, John 25, 240

Dowlands 19, 28-31, 33, 47, 49, 51, 55-56, 58, 60, 63, 101, 106, 132-133, 173, 191, 226, 230

Drake, Martin 165, 185, 188-189

Draper, Jo ix, 20, 54

Dunster, Daniel 49

E

East Cliff Cottage 76, 159

East Devon 16, 30, 56, 87, 99, 163, 167, 196, 223, 235, 238, 240

East Devon AONB 166-167, 196, 235

Ecology iii, 25, 29-30, 33, 36, 95, 113, 143, 147, 192, 235

Ecosystem 116, 121, 218-219, 221, 240, 242-243

Edmonds, Richard 59-60, 62

Edwards, Mike 134, 137-140, 167, 171

Edwards, Rev. Z I 107

Edwards, Sam 18, 75

English Nature   vii, ix, 9–10, 74, 96–99, 110, 116, 119, 163, 169, 189, 196, 213, 224

Erle, T   2

Exeter University   68–69

F

Faults   11, 70–71, 106

Ferns   17, 81, 106–107, 144, 153, 186, 197

Finger and Thumb   82, 91, 233

Flies. See *Diptera (two-winged flies)*

Flints   i, 8, 14, 196, 203

Flora   36, 144, 150–151

Flushes   31, 95, 108, 143, 145, 155, 164

Food chain/web   112, 124, 242

Foot family dinner   52–55

Ford, E B   28

Forestry   ii, 13, 27, 30, 156

Fossils   iv, 8, 14–15, 21, 72, 90, 153–154, 201, 203–204, 223, 239

Fowles, John   vii, 8–9, 13, 15, 38, 47, 56, 90, 93, 125, 201

Foxmould   47, 49, 154, 232

Franks, Elaine   9, 20, 92–93, 140

Freeman, Clare   147

Fungi   viii, 34–35, 37, 108–109, 112–113, 115, 163–165, 179, 186–187, 198, 226, 240

G

Gallois, Ramues   59

Gapper family   19–20

Gentian, Autumn   12, 115, 151–152, 229

Gentian, Early   33, 96, 147, 150, 158, 226

Geology   i–iv, vii, 9, 12, 16, 27, 30, 34, 36, 41, 44, 47, 55–56, 59, 69–70, 72, 82, 85–86, 90, 94–96, 98–99, 116, 144–145, 153, 155, 168, 170, 224, 235

Geomorphology   iii, 23, 58, 79, 155, 168, 170, 182, 224

Gibbs, David   137, 140, 143, 145, 149, 151, 171–172

Gilbert, Oliver   112–114

Goat Island   iii, vii, 21–22, 40, 44, 56, 58–60, 62–63, 66–67, 71, 74, 82, 86, 88, 93, 97–98, 100–101, 106, 109–110, 117, 133–134, 146–147, 149–152, 158, 160, 164–167, 171, 174, 177, 183, 187, 191–192, 194, 198–199, 203, 228–230

Gollop, Ken   viii, 206

Goyle   12, 160

Grainger, Peter   16, 68–69, 72, 74, 78, 225

Grasses   13, 31, 55, 86, 94, 97, 99, 104, 106, 115, 118, 121, 132, 145, 149–150, 152, 155, 158, 167, 172, 186, 189, 243

Gray, Todd   4

Greensand   6, 11, 16, 29, 59, 64, 69–72, 74, 76, 81, 154–155, 203

H

Haeckel, Ernst   236

Hallett, John   2, 48, 65, 161

Hankey, Elizabeth   79, 166

Harvest Mouse   167

Harvestmen   111, 125, 183, 185

Harvey, Prof. L A   33, 37

Haven Cliff   1–4, 22–23, 25, 40, 71, 74, 76–78, 81–82, 87, 91–92, 102, 106, 109–110, 125, 131, 133–134, 137, 139, 159, 178–179, 182, 188–191, 193, 203, 226–230, 233

Hawksworth, David   113–114

Hayward, Peter   215

Hazel Grove   102, 149, 164

Hemiptera (bugs)   172, 185

Henwood, Barry   179, 183, 185

Holm Oak   11, 13, 80–81, 92, 94, 104–105, 108, 115, 117, 141, 145–146, 153, 155, 157, 165–166, 170, 179, 196–197, 201, 224–225, 228, 230

Horsetails   22, 142, 145, 197, 202

Humble Glades   17, 107, 153–154, 156, 158, 170, 174, 179, 184, 192, 228–229, 232–234

Humble Green   153, 230

Humble Point   19, 30, 69–71, 74, 76–77, 104, 113, 146, 153–154, 172, 179, 197, 207, 211

Humble Pond   110, 117, 153–155, 164, 167, 171, 182, 197, 228, 232

Humble Rocks   5, 209, 211–213

Humboldt, Alexander von   235–237

Hutchinson, Peter Orlando   46–47

Hymenoptera   33, 37, 125, 129, 137, 172, 185

I

Ichthyosaurs   9, 14–15

Insecticides   124, 240–242

Insects   93, 95, 129–130, 137, 151, 158, 168, 179, 183, 192, 194, 240

Invertebrate Conservation Trust. See *Buglife*

Invertebrates   11, 11, 22, 25, 28, 96, 111, 115, 117, 119, 125, 130, 134, 137, 143, 156, 167–169, 171–173, 226, 234, 243

J

Jackson, Peter   99

Jefferies, Roy   108, 145, 152, 163

Jurassic   iv, viii, 14, 16, 29, 66, 70–72, 94, 235, 239

K

Keble Martin, W   33

Kennard, Jim   88, 91, 93–94, 98, 116, 119

Kilvert, Francis   2–3

Knott, Albert   vii, 79, 97, 99, 116, 118, 153, 162–163, 213, 224

L

Landslip   i–ii, 1, 10–11, 16–17, 21–23, 28–31, 43–44, 47–49, 51–52, 54–56, 58–61, 64, 67, 69, 71–72, 74, 76, 82, 86–

87, 94-95, 101, 108, 114, 116, 122, 130, 149, 154, 158, 160, 168, 182-183, 192, 197-198, 202, 208, 226, 233-234, 240

Landslip Cottage 19-20, 32, 57, 88, 198

Langham, Peter & Bob 15

Laurel, Cherry 11, 159, 170, 182-183, 189

Leland, John viii, 1, 4

Lepidoptera. See *Butterflies & butterfly conservation, Moths*

Lias iv, 6, 8, 11, 14, 27, 29, 43, 47, 59, 71, 74, 76, 88, 202, 204, 206-207, 211

Lichens viii, 110-114, 118, 163-165, 186, 212, 226, 234

Limestone 7, 15, 19, 95, 113, 145, 153, 155-157, 201, 206-207, 211-212, 232

Lithograph 14, 49

Littoral 113, 182, 211-212

Liverworts 34, 108, 118, 131, 144, 153, 164, 183, 186

Lizards 160, 167, 184, 186

Lobster 18, 56, 160, 207, 209, 218, 221

Lock, Mike 144, 150, 183

Lyme Regis vii-viii, 1, 4-7, 9, 12-17, 22, 27-28, 32, 34, 45, 51, 56, 58, 72, 74, 77, 82, 88, 90, 99, 107-108, 114, 130, 163, 169, 174, 194, 198, 201-202, 204, 206-207, 212, 217, 219-221, 226, 228, 235, 243

Lynch Cottage 12, 122

Lynch Meadow 156, 228

M

MacFadyen, Dr 12, 74, 76

Mammals 167, 240

Management Plans vii, 85-86, 99, 101, 103, 105, 115-116, 121, 143, 224-225, 234

Marine Biological Association 129

Marine Protected Area (MPA) 25, 95, 220-221

Marl 6, 23, 72

Marren, Peter 150

Marsh, George Perkins 236

Matthews, Fiona 187

May, Harry 203

McCormick, Roy 118, 165

McGowan, Chris 9, 15

Merrett, D P 37, 125-126

Micraster 203-204

Millipedes 133, 185, 242-243

Molluscs

  Marine 212-213, 217

  Terrestrial 130-132, 154

Monmouth Beach 204

Moore, Keith 13, 19, 41, 59, 69, 79, 99

Moore, Norman 32, 36, 96, 121, 124, 131, 241-242

Morrison, Herbert 26

Mosses viii, 34, 81, 108–110, 117–118, 131, 138, 144, 152, 164, 183, 186

Moths 32, 39, 111–112, 118, 151, 156–157, 159, 165, 172, 179, 183, 185, 187, 191–192, 194, 226, 241

Mowing 93, 97–98, 115, 149, 228

Muir, John ix, 236–238

Museum 27

  British 114

  Exeter 32, 127

  Lyme Regis (Philpot) vii, ix, 9, 14–15, 20, 44–45, 47, 54, 56, 90–91, 107, 163, 206, 224

  Oxford 47

  Victoria & Albert 6

N

National Nature Reserves ii–iv, 12, 25–26, 31, 34, 56, 58, 85, 91, 104, 121, 130, 147, 156, 163, 170, 182, 226, 234–235, 240

National Parks iii, 25, 238, 240

Nationally [very] Scarce 119, 134, 137–138, 140, 157, 171–172, 179

Natural England ix, 59, 81–82, 118, 182, 228–229, 233–234

Nature Conservancy (& NC Council) 12, 26, 32–34, 36, 56, 85, 87, 89, 93–94, 98, 101–102, 113, 125, 156, 241

Nature Reserves Committee 25

Naylor, Paul ix, 215

Neate, George 183

Newts 93, 167, 228

Nicholson, E M (Max) 25–26, 33, 85

Nightingale 21, 95, 122, 124

Nottingham Catchfly 96, 114, 118, 166, 226–227, 232

O

Opilionids. See *Harvestmen*

Orchids 10, 22, 32, 38–40, 93, 106, 118, 121, 143–145, 150–152, 158, 160, 163, 198, 229, 232

Ornithology. See *BTO (British Trust for Ornithology)*

Orthoptera 134, 139, 185

Orton, P D 34–35

P

Page, Phil 97–98, 118, 151, 171, 198, 224

Palaeontology iv, 8, 29, 94, 117

Palmer, Dave 79

Parasites 127–128, 130, 155, 177, 215, 236

Parr, Phil ix, 108, 118, 141, 144, 156, 173, 177

Peek family viii, 12, 17–18, 88, 107, 169

Peregrine 33, 81, 95, 101, 117, 122, 124–125, 134, 174, 176, 186

Pesticides 124, 241–242

Pheromone 151, 159, 179

Picturesque 4, 43, 47, 51

Piddock 183, 201, 206, 212

Pinhay 4, 10-15, 17, 19-20, 27, 29-31, 34, 58, 69, 71-72, 74, 76-79, 82, 92, 102, 106, 108, 113-114, 118, 121, 139, 143, 154, 156, 159, 161, 172-173, 182, 191, 194, 196, 201-203, 207, 211-212, 225, 228-229, 232, 234, 240

Pitfall traps 125-126, 183

Pitts, John 23, 59-60, 74, 76-77, 82, 87, 210

Plateau, the vii, 19, 33, 40, 66, 70, 93, 95, 97-98, 114-117, 124, 133, 141, 146-150, 158-159, 165, 172, 174, 177, 179, 191-192, 198, 233, 240

Plesiosaur 14

Plymouth University 62, 220

Pole, Sir W 51

Pool, Mark 108, 152, 163, 186

Prawn 56, 207-209

Pritchard, Laurie 12, 56

Proctor, M C F 34, 37, 108

Pulman's 20

Pumping Station 11-13, 16-18, 34, 58, 68-69, 71-72, 74, 76, 78-79, 81, 93, 104, 106, 116, 166, 196, 203, 225-226, 228

Q

Quadrats 36, 146-147, 149, 151-152, 225

R

Rabbits 48, 58, 64, 86, 97, 117, 121, 167, 240

Radiolarians 236

Raine, Brian 177

Rainfall 14, 72-74, 78, 161

Randall, John 185

Ratcliffe, Derek 124

Ravine Pond 10, 104, 195-196

Red Data Book (RDB) 131, 134, 137-138, 140, 157-159, 168, 172-173, 189-191, 194

Reptiles iv, 14-15, 167

Resting Stone 19, 21

Rhododendron 106, 169

Roberts, George 6-7, 12, 97

Rockpooling 206

Rogers, David 94, 97-99

Rose, Sam 171

Rousdon viii, 12, 17, 19-20, 27, 31, 34, 58, 69, 71-72, 74, 76-78, 109-110, 131, 169, 174, 179, 225

Rudge, Doug 154, 234

Rushes 22, 32, 145, 171

S

Sabellaria 201, 212-213

Sandstone 1, 44, 102-103, 113, 143, 160

Scallops 203, 217–221

Sea Anemones 183, 213, 215–217, 219–221

Sea Fan 183, 218–221

Seaton viii, 1, 4, 12, 17, 19, 22, 27–28, 55, 63, 87, 93, 102, 122, 169–170, 198, 208, 219

Seaweeds 43, 113, 170, 183, 201, 206, 211–213

Seepage 16, 23, 72, 95, 134, 143, 172, 189, 192

Seven Rock Point 14

Sheehan, Emma 220

Sheep-wash 20–21, 99, 160, 195–196, 198, 227–228

Silica 203

Slabs, the 88, 159

Snails 102, 130–132, 186, 191

Soft Cliffs 97, 109, 137, 143, 168–169

South West Water 68–69, 72, 74, 116, 225–226

Special Area of Conservation (SAC) iii, 97, 117, 143, 168, 189, 226

Spiders 37, 111, 125–126, 130, 138–139, 156, 168, 171–172, 177, 183, 185

Sponges 183, 213–214, 216–217, 219, 221

Spooner, Malcolm 33, 37, 125, 127, 129

SSSI iii, 99, 230

Stamp, Sir Dudley 27

Stedcombe 56, 86

Stoneboats 207

Stratotype 72

Stuart Line Cruises 82, 169

Stubbs, Alan 93, 119, 190, 192

Students 27, 29, 34, 36–37, 39, 69, 79, 97, 99, 107–109, 165–166

Succession iii, vii, 9, 16, 27, 70, 72, 94–95, 167, 169, 203, 226, 243

Sunderland, Tom iv, ix, 22, 79, 153, 159, 182–183, 232

Sutton, Stephen 112, 127–128, 171

Swete, John 4, 6

T

Tansley, Sir Arthur 25, 27–28, 30, 33, 85

Taylor, Sir William 30

Tetrads 33, 141, 164, 174

Thoreau, Henry David 236–237

Tithe Map 16, 197

Transect 80–81, 121, 147–148, 166, 212

Trees 17, 19, 29, 37, 81–82, 85, 94, 101–102, 104–106, 112–115, 117, 134, 141, 144–147, 152–156, 163, 166–167, 170, 179, 183, 186, 191, 193–198, 230, 233, 235, 240, 243

Triassic iv, 16, 70, 72, 161, 222–223, 239

Turner, Adrian 131

U

Umble Rock. See *Humble Rocks*

Unconformity iv, 62, 70–71

Underhill Farm vii, 8–11, 31, 38, 58, 91, 102, 194, 235

University College, London 32, 37, 126, 147

W

Wallace, Tom vii, 10, 13, 33–35, 37, 39, 56, 59, 74, 76, 85, 87–88, 96–97, 108–109, 116–118, 122, 125, 131–132, 134, 138–139, 141, 164, 212, 228

Ward, B T 31

Wardens 10, 12, 56, 87–89, 91, 93–94, 96–98, 101, 114, 116, 122, 147, 219

Ware 10, 29, 31, 33, 58, 74, 77, 93, 99, 106–107, 114, 118, 149, 156, 159, 177, 191, 203

Waters, Marjorie 146, 154, 177, 179

Watson, L J 27–28

West Cliff Cottage 16, 75, 106, 196–197, 240

White, Walter 6–7, 55

Whitechapel Rock 15

Whitehouse, Andrew 137

Whitlands 5, 11, 16–17, 29, 44–45, 49, 74, 76, 102, 104, 107, 109–110, 113, 141, 147, 153–154, 157, 165–166, 179, 191, 195–197, 203, 240

Wildlife and Countryside Act 189, 218

Winefred 1, 170

Wolton, Rob 94, 185, 189

Woodland, John 133

Woodlice 112, 119, 126–128, 133, 171, 184–185

Woodruff, Chris 167

Woodward, H B 74, 76, 82, 250

World Heritage Site iii, viii, 9, 59, 69–70, 91, 98, 169, 171, 223, 226, 238

Worms

Marine 201, 212–213, 217–219

Terrestrial 130, 160, 167, 186

Wulf, Andrea 235

Y

Yellowstone National Park 237

Yosemite National Park 237

Youngman, Peter 79, 167

Image credits

Abbreviations:

BL - Buglife
BTO – British Trust for Ornithology
CS – Creative Solutions, Axminster
DA – David Allen
DC – Donald Campbell (Author)
DCM – Dorset County Museum
EF – Elaine Franks
EN – English Nature
Enc – Encyclopedia of British Birds
FSC - Field Studies Council
JKSJ - J K St Joseph
JM – John Marriage
JP – John Pitts
LRM – Lyme Regis Philpot Museum
MA – Muriel Arber
ML – Mike Lock
NC – Nature Conservancy
NCC – Nature Conservancy Council
PB – Paul Butter
PP – Phil Parr
RE – Richard Edmonds
UCL – University College London
V&A – Victoria & Albert Museum
Woody – Richard Matthews

Page	Credit
Cover	JM
Map	NC
v	'Raised reefs occasioned by the landslip'. Drawn and published by J. Baker
2	JKSJ (Cambridge University 1949)
3	Probably PP
5	C W Bampfylde (V&A)
6	C W Bampfylde (V&A)
7	Rev. John Swete
8	LRM
10	Unknown
11	LRM
13	NCC
14	JM (both)
15	Pat Bennett
16	George Cumberland (above), LRM (below)
18	LRM (both)
19	Keith Moore redrawn by Val Baker
22	Woody (both)
23	Roger (or Kath) Critchard
28	Portrait by L J Watson, Dowlands Slip - JKSJ
31	ML (both)
32	Orchid – DA, Stonechat - Enc
33	BTO

34	DA
35	DC
36	PP
37	PP, Unknown (below)
39	Tom Jenkyn
40	PP
42	'A Gentleman'
45	G Hawkins (above), LRM (below)
46	P O Hutchinson (above), Anon in the *Penny Magazine* (middle), *Oxford Today* (below)
50	LRM
51	LRM
52-53	W Porcher (DCM)
54	W Porcher (DCM)
57	DC (top), LRM, DC (middle), MA (bottom)
60	Searle (top), JP (middle), RE (bottom)
61	William Dawson / William Conybeare; CS (after Dawson)
62	RE
65	CS, William Barnes
67	Anon (both) – LRM
68	From Campbell (2006)
70	Nomination Handbook
75	Norman Lambert collection
77	NCC
78	DC
80	Liz Hankey (all)
82	JP
83	H T de la Beche (top), H B Woodward (middle), A J Jukes Brown (below)
87	JP
88	MA
89	Margaret Pogson (above), PP (below)
90	Sent by Sarah Fowles
92	EF
97	JKSJ
98	NCC (above), DC (below)
99	PP
100	Val Baker after Norman Barns
103	NCC (top), DC (left), NC (right)
107	From Rev. Z I Edwards
111	Fungi – artist unknown
112	Val Baker after Oliver Gilbert
114	Enc
115	DC
116	John Muir Trust
120	Buzzard – Donald Watson, Fox – PP, Barn Owl & Partridge – Enc, Minotaur Beetle – Norman Moore
122	Laurie Pritchard
123	All Enc
127	Hilary Burn
128	Val Baker after Hilary Burn
129	FSC/BL
133	Peter Vernon
135	Enc (both)
136	Enc (all)
140	EF
141	PP
142	JP
144	'SH' (Wiltshire Naturalists Trust)
146	PP (both)
147	PP (both)
148	UCL (top), NCC (left), J Jesty (right)
150	PP (all)
151	PP
152	PP (top), DC (below)
155	PP
157	PB (above), Ron Bragg (below)
159	NCC (left), Hilary Burn (right)

160	DC (top left), NC (bottom right)	212	DC
162	PP (all)	213-219	All Paul Naylor
165	S Luscombe	220	Devon Wildlife Trust
166	JKSJ (top), DC (below)	222	DC (all)
167	John Muir Trust	227	FSC / Butterfly Conservation (top right), DC (remainder)
168	Kevin Page (EN)		
169	World Heritage Coast Trust / EN	229	PP (both)
170	John Seward (above), Sam Rose (below)	231	Woody (all)
		234	JM
171	DC (above), Hilary Burn (below)	237	Library of Congress
175	Enc (all)	239	DC (all)
177	The Times	242	https://www.flickr.com/photos/49399018@N00/49474085591/
178	PP		
179	PP (all)		
180	PP (all)		
181	PP		
184	PP (both)		
187	Enc (above), Tom Wallace (below)		
188	Tony Todd		
193	PP (lower right), PB (the remainder)		
194	DC		
195	PP (lower left), DC (remainder)		
196	DC (top), PP		
197	DC, Tithe Map redrawn by Val Baker (below)		
199	DC (top), FSC (middle and bottom)		
200	NCC		
201	Enc		
202	DC (both)		
205	DC (Predator), Enc (rest)		
204	JM (above left), Colin Dawes (below left), Nikko Tinbergen (below right)		
206	Colin Dawes (above), DC (below)		
207	LRM (all)		
208	Margaret Pogson		
209	FSC		
210	JP (top), CS (bottom)		